Ecological Studies

Analysis and Synthesis

Edited by
W.D. Billings, Durham (USA) F. Golley, Athens (USA)
O.L. Lange, Würzburg (FRG) J.S. Olson, Oak Ridge (USA)
H. Remmert, Marburg (FRG)

Volume 59

Ecological Studies

J.O. Reuss
D.W. Johnson

Acid Deposition and the Acidification of Soils and Waters

With 37 Illustrations

Springer-Verlag
New York Berlin Heidelberg Tokyo

J. O. Reuss
Department of Agronomy
Colorado State University
Fort Collins, CO 80523
U.S.A.

D. W. Johnson
Environmental Sciences Division
Oak Ridge National Laboratory
Oak Ridge, TN 37831
U.S.A.

Library of Congress Cataloging-in-Publication Data
Reuss, J.O.
    Acid deposition and the acidification of soils and waters.
    (Ecological studies ; v. 59)
    Includes index.
    1. Soil acidification. 2. Acid deposition—
Environmental aspects. I. Johnson, D.W. (Dale W.),
1946–   . II. Title. III. Series.
S593.5.R48 1986   631.4'2   86-3805

Although the research described in this article has been funded wholly or in part by the United States Environmental Protection Agency (EPA) through Interagency Agreement Number 40-740-78 to the U.S. Department of Energy, it has not been subjected to EPA review and therefore does not necessarily reflect the views of EPA and no official endorsement should be inferred.

Typeset by Bi-Comp Inc., York, Pennsylvania.
Printed and bound by Halliday Lithograph, West Hanover, Massachusetts.
Printed in the United States of America.

9 8 7 6 5 4 3 2 1

ISBN 0-387-96290-5 Springer-Verlag New York Berlin Heidelberg Tokyo
ISBN 3-540-96290-5 Springer-Verlag Berlin Heidelberg New York Tokyo

Preface

The majority of this book was written in 1983–84 while the senior author was a Visiting Scientist at Oak Ridge National Laboratory (ORNL) in Oak Ridge, Tennessee. We believe that the approach to the problem of acid deposition effects on soils and waters developed during this collaboration contains elements that are significantly different from most prior work in this area. Some of the material and the software used in the development of these concepts stem from earlier individual efforts of the authors. However, what we believe to be the more significant concepts concerning the processes by which alkalinity may be developed in acid soil solutions, and by which acid deposition may contribute to the loss of this alkalinity, were the result of this collaboration.

The ultimate usefulness of these concepts in understanding and dealing with various aspects of the problems associated with acid deposition cannot be adequately gauged at the present time. They must first withstand tests of consistency with available observation, and of direct experimentation. It is our hope that dissemination through this book will facilitate this process within the scientific community.

The authors wish to thank the administration of the Environmental Science Division at ORNL, and the College of Agricultural Sciences at Colorado State University for their support in arranging this collaboration. We also wish to express our appreciation for the financial support provided by EPA.

Personal thanks are due to Dr. Jerry Olson of ORNL who originally suggested that the material might be appropriate for publication in the Ecological

Studies Series, and who made valuable suggestions concerning the manuscript. We also wish to thank Dr. James Galloway of the University of Virginia at Charlottesville, Dr. Robert S. Turner of ORNL, and Dr. Bernard Ulrich of the University of Göttingen, Federal Republic of Germany, for their efforts in reviewing the manuscript and providing valuable suggestions.

    This research was funded in part by EPA/NCSU Acid Precipitation program [a cooperative agreement between the USEPA and the North Carolina State University (NCSU)], and a subsequent agreement (AP-0307-1983) between NCSU and Colorado State University, and in part by the National Acid Precipitation Assessment Program of the U.S. Environmental Protection Agency under Interagency Agreement 40-740-78 with the Office of Health and Environmental Research, U.S. Department of Energy, under Contract No. DE-AC05-840R21400 with Martin Marietta Energy Systems, Inc.

J.O. Reuss
D.W. Johnson

Contents

1. Introduction

One of the most pressing issues currently facing environmental scientists is the need to predict the effect of acid deposition on terrestrial and aquatic ecosystems. As is typical of issues concerning public policy, a great deal of controversy surrounds the discussion of both the nature and extent of acid disposition effects, as might be expected where decisions of great economic import must be made on the basis of necessarily limited scientific knowledge. Here, the authors attempt to analyze the effect of acid deposition (1) on the soil–plant system and (2) on the composition of the solution that is released to surface waters and groundwaters. In so doing, the linkage between effects on terrestrial and aquatic systems is elucidated. Although it is unrealistic to expect that such clarification will materially lessen the controversy surrounding the issue, perhaps it will help to focus scientific investigation on those processes most likely to control the effects of acid deposition on soils and waters.

1.1 Rationale

Soil–plant systems are extremely complex. The scientist who attempts to evaluate the effects of acid deposition on these systems is soon confronted with a maze of chemical reactions and biological processes involving H^+ ion (proton) transfers. The conclusions drawn by various researchers may be very different, depending on the perspective from which the problem was viewed and/or the experimental methods used. To provide a conceptual framework for evaluation of the probable long-term effects of acid deposition, we have tried to focus on a few fundamental soil processes. After a brief overview, individual chapters address the specific topics of soil acidification, sulfate interactions, nitrogen effects, soil-solution equilibria, canopy interactions, the aquatic interface, and soil and water sensitivity. The final chapter (Chap. 9) presents a synthesis of the concepts developed in previous chapters and a brief discussion of how these perceptions of acid deposition effects may differ from currently accepted concepts.

Our intent is neither to compile an exhaustive review of the literature on the subject nor to discuss opposing viewpoints in detail, but rather to provide a conceptual model that is consistent with established physicochemical principles and the bulk of available information. One of the most important tools used in the development of this conceptual model is a rather simple chemical equilibrium model (documented in Appendix A). We have found that such a model provides insights into interactions among chemical processes that are not otherwise apparent. A reasonably clear conceptual picture emerges and is summarized in Chap. 9. It is our hope that it will be useful in (1) providing insight into probable long-term effects of acid deposition, (2) formulating testable hypotheses, and (3) guiding the design and interpretation of research.

For the most part, our discussion focuses on those systems in which acid deposition is most widespread and deleterious effects have been observed. These tend to be forested ecosystems in which rainfall exceeds evapotranspiration for at least part of the year, so that percolation of solutes below the root zone is a normal occurrence. We shall exclude soils containing free carbonates, those formed predominantly under reducing conditions as a result of flooding, and heavily managed systems in which acidification pressures from such practices as nitrogen fertilization generally far exceed those resulting from acid deposition.

1.2 An Overview

Acid atmospheric deposition enters an ecosystem in a variety of forms and via a number of pathways (Fig. 1.1). The most common strong acids in acid precipitation are sulfuric (H_2SO_4) and nitric (HNO_3). Occasionally other mineral acids [e.g., hydrochloric (HCl) or phosphoric (H_3PO_4)] or organic acids [e.g., formic (HCOOH) or acetic (CH_3COOH)] may be present, although not generally in significant quantities. Although a substantial amount may be present, the Cl^- anion generally derives from neutral salts, usually of marine origin, rather than from HCl (Henriksen 1979, 1980). Other organic acids are also present as is carbonic acid (H_2CO_3), formed by the reaction of atmospheric carbon dioxide (CO_2) with water. The nature of this latter reaction is such that the contribution of carbonic acid to the acidity of the solution decreases as the contribution of strong acids increases. Although carbonic acid is an important acidifying agent in natural systems, the amount in soil solution is generally determined by the

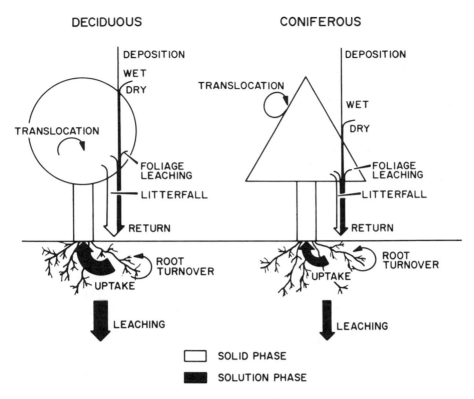

Figure 1.1. An overview of acid deposition on a forest ecosystem.

CO_2 content of the soil gases (which is usually much higher than in the atmosphere), so that the amount that actually enters via rainfall is irrelevant.

Although H^+ is usually the dominant cation in acid deposition, other cations may accompany the acid anions. Among those commonly encountered are Ca^{2+}, Mg^{2+}, Na^+, K^+, and NH_4^+. Thus, acid precipitation is usually a complex mixture of acids and salts. The degree to which the strong acid anions (SO_4^{2-}, NO_3^-, and Cl^-) exceed the basic cations, on a chemical equivalent basis, determines the capacity of precipitation to acidify the system. An important exception to this is the NH_4^+ ion, which can actually enhance the capacity of precipitation to acidify the system (Chap. 4).

Atmospheric deposition includes not only wet precipitation but gaseous and dry particulate components as well. Important gaseous components include sulfur oxides (SO_x), particularly SO_2, nitrogen oxides (NO_x) and ammonia (NH_3). These components may be adsorbed on soil or plant surfaces or absorbed within the plants through the stomata and are eventually converted to H_2SO_4 and HNO_3. Ammonia gas may also be deposited and converted to NH_4^+. Particulates deposited may contain salts of any of the acid anions or basic cations. The total acidifying capacity associated with these sources often equals and may exceed that of the direct acidity of the precipitation.

When acid precipitation impinges on the system it may either enter the soil directly or interact with the canopy. Within the canopy, it may become more acidic as the result of washoff of accumulated dry deposition or by leaching organic acid anions from the foliage, and of leaching of the reaction products of dry-deposited SO_2. Acid precipitation, on the other hand, may become less acidic because of processes that may be thought of as exchanging H^+ for basic cations in the plant tissue. Of these processes, only the washoff of dry deposition represents a net change in the acidity of the soil–plant system. The other changes affect internal cycling only, although they may affect the vegetation by prematurely removing cations from active tissue.

On entering the soil, solutions undergo many reactions. When deposition dominated by sulfur enters a system previously unaffected by acid deposition, the sulfate (SO_4^{2-}) concentration of the soil solution begins to increase. Some soils have substantial capacity to adsorb SO_4^{2-} on the particle surfaces, and in such cases, this adsorption acts as a buffer, delaying the elevation of solution SO_4^{2-} concentrations. The SO_4^{2-} uptake and cycling capacities of the biotic component of the ecosystem also act as buffers, slowing the increase in solution SO_4^{2-} concentration. Usually, the soil system will reach an equilibrium at which elevated SO_4^{2-} concentrations in solution become sufficiently high that the outgoing flux of SO_4^{2-} sulfur in the drainage water is approximately equal to the incoming sulfur in acid deposition. Because soils vary greatly in adsorption capacities, the time required before substantially elevated SO_4^{2-} concentrations are observed in the drainage waters may vary from a few weeks to many decades.

As SO_4^{2-} concentrations in soil solutions and water leaving the system increase, charge balance considerations dictate that these anions be accompanied by an equivalent amount of cations. In soils that are even moderately well supplied with bases (i.e., perhaps 15% or more of the negatively charged exchange sites are occupied by the basic cations Ca^{2+}, Mg^{2+}, K^+, or Na^+), these basic cations will comprise most of the increase in solution cation concentration. The remainder will be largely H^+ or aluminum species, particularly Al^{3+}, that comes into solution as a result of exchange reactions or the dissolution of soil minerals, although in some systems iron or manganese species may be significant. The increase in the rate of removal of basic cations will tend to acidify the system. Because the total supply of these cations in the soil is usually quite large relative to the annual input of H^+ in acid deposition, acidification of soils and waters by this cation export mechanism is likely to be a slow process involving decades or even centuries in deep soils or soils containing significant amounts of minerals that release basic cations upon weathering.

In low-base-status soils, a significant fraction of the increased cations in solution and in discharge waters may consist of H^+ and ionic aluminum species. This is generally undesirable because the Al^{3+} ion is toxic to many species of plant and animal life and because both aluminum and H^+ will reduce the alkalinity of the discharge water. In some cases this reduced alkalinity and increased aluminum content may be sufficient to cause drastic changes in the aquatic biota and loss of fisheries.

The change in solution composition brought about by increased SO_4^{2-} concentrations resulting from acid deposition is mediated by a complex set of reactions with the soil system and is not simply a "wash through" of the acid impinging on the system. The effect is not only an increase in the cation concentration in solution but a change in proportions of the various cations as well. An understanding of what the distribution of cations is likely to be following changes in rainfall chemistry is crucial to our ability to predict the effects of changes in acid deposition. Therefore, the ensuing chapters focus on the processes that determine the distribution of ions in solution and the implication of changes in acid deposition loading on the composition of the soil solution and surface water.

Other processes are, of course, likely to be important in assessing the overall effects of acid deposition. Atmospheric deposition contains sulfur and nitrogen, both of which are essential plant nutrients. The fertilization effect of nitrogen in particular may be substantial, so these and other processes are discussed as well.

Some processes that may be significant are not discussed, either because of a lack of information on the degree to which they affect the chemistry of the system or because of a lack of competence of the authors in some specialized areas. One such process is what is commonly termed "macropore" flow. Obviously, some fraction of the impinging rainfall may enter the surface or subsurface waters directly, interacting with the soils or the vegetation only minimally. The degree to which such nonequilibrium transport occurs will be very different among systems. Unfortunately, we cannot contribute very much at this time to the understanding of the importance of this process to the effect of acid deposition on soils and waters. Another major area not covered is that of the possible effects on soil biological processes. This omission does not stem from any lack of appreciation of the importance of these processes but from a combination of lack of competence in a specialized area and a conviction that meaningful evaluation of effects on these processes must await an improved understanding of the changes likely to occur in the surrounding chemical environment within which the processes take place. Therefore, we have chosen to focus here on changes in the chemical environment.

2. Soil Acidification: Fundamental Concepts

Natural soil acidification processes have been recognized and studied for decades or perhaps centuries. An understanding of these processes is essential to an understanding of soils and of natural and agricultural ecosystems. One of the most important characteristics of soils is the cation-exchange complex. These are negative charges, either on clay minerals or on soil organic matter. In the case of the clay minerals, these charges usually arise from isomorphic substitution within the mineral lattice of a cation of lower positive charge for one of higher charge. In the case of organic matter, the charges arise mainly from the ionization of H^+ from carboxyl, phenol, and enol groups (Coleman and Thomas 1967). In alkaline or neutral soils, the negatively charged exchange complex is dominated by basic cations (i.e., Ca^{2+}, Mg^{2+}, K^+, and Na^+). In acid mineral soils this complex is usually dominated by aluminum species [i.e., Al^{3+}, $Al(OH)^{2+}$, and $Al(OH)_2^+$] formed by the dissolution of soil minerals in acid systems. In acid organic soils, H^+ may be the dominant exchangeable cation. The acidity of a soil is thus determined by the relationship between the amounts of the basic cations and the acid aluminum species on the exchange complex. Processes that would tend to acidify a soil include those that tend to increase the number of negative charges, such as organic matter accumulation or clay formation, or those that remove basic cations, such as leaching of bases in association with an acid anion. Processes that would tend to make a soil more basic would add basic cations, either from outside sources or from the weathering of soil minerals, or reduce negative charge, such as might occur during the destruction of organic matter by fire.

Very acid soils are usually characterized by relatively high levels of aluminum species, particularly the Al^{3+} ion, in solution. This ion is toxic to some plants at concentrations as low as a few milligrams per liter, whereas other plants can tolerate tens or even hundreds of milligrams per liter with no apparent ill effects. For instance, McCormick and Steiner (1978) found that hybrid poplar showed toxicity symptoms at solution aluminum concentrations of 10 ppm, whereas black alder, birch, pines, and oaks tolerated aluminum concentrations of 80 to 160 ppm. Tolerance to Al^{3+} also varies with the concentration of other ions in solution, particularly Ca^{2+} and PO_4^{3-} (Ulrich 1983). Many of the deleterious effects that have been traditionally ascribed to soil acidity are now recognized to result from aluminum toxicity. In terms of the effect of acid precipitation on the system, base removal and aluminum mobilization can be identified as key processes of concern, while recognizing that the effects of these processes may be modified by species differences and solution composition, as well as by other factors such as mineral weathering and changes in organic matter content.

The term "acidification," as it has been applied to soil systems, actually refers to a complex set of processes. It cannot be quantitatively described by any single index. One of the most useful concepts in this regard is that of capacity and intensity factors. Capacity refers to the total storage of one or more components and, in terms of soil acidity, would usually refer to the storage of protons or Al^{3+} (although not necessarily in ionic form) or to the storage of base cations on the ion-exchange complex or in weatherable minerals. Intensity refers to the concentration in solution at any one time or, in the case of H^+, the solution pH. In terms of capacity factors, the most likely effect of acid deposition is an increase in the exchange acidity and a reduction of the exchangeable bases (Fig. 2.1). The exchange acidity is increased either directly by the H^+ inputs or, more likely, by increasing exchangeable aluminum through the reaction of H^+ with soil minerals. The reduction in exchangeable bases occurs through the replacement of base cations on the exchange complex by aluminum species. The base cations are then leached from the system in association with the strong acid anions (SO_4^{2-} or NO_3^-) from acid deposition. Although some reported deposition rates, particularly in evergreen spruce forests (Matzner 1983; Höfken 1983), are above those assumed in Fig. 2.1, it is apparent that total exchange acidity is far in excess of annual acid inputs. Therefore, substantial changes in total exchange acidity resulting from acid deposition would not be expected unless the system were dramatically affected for many decades. In acid soils, the exchangeable base pool would be substantially smaller than the total exchange acidity, and under heavy deposition loading, significant losses could take place within perhaps a couple of decades. Measurements of soil acidification rates, whether the result of natural internal processes or of anthropogenic influences, often involve the estimation of ion fluxes into or out of these "capacity" pools (Matzner 1983; Matzner and Ulrich 1983; van Breemen and Jordens 1983).

Although ion flux measurements provide a useful frame of reference for comparing proton fluxes in natural systems with those affected by acid deposi-

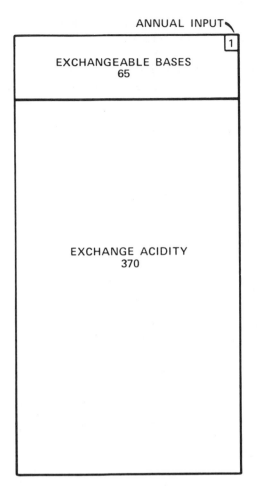

Figure 2.1. Typical pool size of exchangeable bases and exchange acidity (H^+ + alumi-num species) relative to annual H^+ input from acid deposition. Assumptions include an input of 1000 mm pH 4.2 rainfall plus an equal amount of acidity in dry deposition to 300 mm soil with bulk density 1.2, cation exchange capacity (CEC) of 0.15 eq/kg, and 15% base saturation.

tion, they are not particularly helpful in estimating intensity effects (i.e., pH changes in soil solutions or surface waters). These intensity changes are partic-ularly important because of their relationship to such important effects as alu-minum mobilization and surface-water acidity. The charge balance principle requires that any increase in anion concentration in soil solution that occurs as a result of acid deposition must be accompanied by an equivalent increase in cations (see Chap. 3). The physicochemical relationships governing soil-solu-tion equilibria (Chap. 5) are such that increased solution concentrations will increase the proportion of cations of higher valence versus those of lower valence (i.e., Al^{3+} will increase relative to Ca^{2+} and Mg^{2+}, whereas the proportion of these dialent ions will increase relative to H^+, K^+, and Na^+). It is difficult to overemphasize the importance of these "intensity factors" because they determine the changes in solution concentration that are likely to occur as a result of acid deposition. These factors are discussed in considerable detail in the ensuing chapters.

If the conceptual model is accepted in which a reduction in the pool of exchangeable bases results in acidification, it follows that the net accumulation of basic cations in forest biomass is intrinsically acidifying unless these cations are replaced by outside inputs or the release of base cations through weathering of soil minerals. Alban (1982), for example, found that in Minnesota soils beneath 40-year-old aspen and white spruce were lower in exchangeable Ca^{2+} and pH (5.3 to 5.6) than soils beneath adjacent 40-year-old jack- and red-pine stands (pH = 6). However, the organic soil horizons under the aspen and spruce stands had higher pH and calcium than the pine stands because of the more rapid calcium cycling in the former. Thus, uptake of bases by plants may result in redistribution of basic cations in the system as well as acidification of at least a part of the soil profile.

Harvest removal of vegetation causes the export of base cations, and regrowth after harvest may cause further acidification of soils, especially in calcium-accumulating species such as *Quercus* and *Carya spp.* (Johnson et al. 1982b). Without harvesting, the acidifying effect of cation accumulation will continue as long as biomass accretion continues. Fire will reverse this accumulation; however, burning can cause surface alkalization without substantially reversing the subsoil acidification caused by tree cation uptake (Grier 1975). It may also release some nutrients by loss in smoke or sudden runoff.

Acidification resulting from plant uptake of base cations is often explained as being caused by release by the plant root of H^+ ions in exchange for base cations or of OH^- (or HCO_3^-) ions in exchange for anions taken up (Reuss 1975, 1977; Miller 1983; Matzner 1983; Matzner and Ulrich 1983). In this model, the net acidification is taken as the difference between the net uptake of cations and anions. Inherently, this approach implies that acidification be defined as an increase in the H^+ ion pool or exchange acidity. It is also often necessary to assume a distribution between nitrogen uptake in the NH_4^+ and NO_3^- forms (Miller 1983; Matzner 1983; Matzner and Ulrich 1983). Because the nitrogen cycle, both within the plant and in the soil, is inherently balanced (Chap. 4), we prefer to consider acidification resulting from base uptake by plants in terms of reduction in the pool of exchangeable bases, at least to the extent that these are not replaced by the weathering of soil minerals. Because this exchangeable base pool is smaller than the exchange acidity, changes in exchangeable bases are more likely to be reflected in the critical intensity factors than are changes in total acidity.

The second major process by which base removal occurs is leaching. The principle of electroneutrality requires a balance of positive and negative charges in solution. Therefore, removal of basic cations in solution can occur only in association with a mobile anion. In most neutral or moderately acidic soil solutions, the dominant anion in solution is HCO_3^-. The activity of HCO_3^- in solution is a function of pH and CO_2 partial pressure such that

$$(H^+)(HCO_3^-) = P_{CO_2} \cdot 10^{-7.81} \qquad (2\text{-}1)$$

where the parentheses denote solution activity and P_{CO_2} refers to the partial pressure of CO_2 in atmospheres. In previous papers, the authors have discussed this relationship in the context of acid precipitation effects on soils (Reuss 1975, 1977, 1978, 1980; Johnson et al. 1977).

Soil acidification under the influence of a strong acid anion such as SO_4^{2-} is rare in natural systems. Some perspective on the relative concentrations of HCO_3^- and SO_4^{2-} ions in natural systems and in acid deposition can be gained from Fig. 2.2. In this figure, any point on the H^+ line above the HCO_3^- line relevant to the controlling CO_2 partial pressure requires the presence of some anion other than HCO_3^-. In acid precipitation, the dominant strong acid anion is commonly sulfate, and the resultant increase in the concentration of this strong acid anion in the soil solution can have a significant effect on solution composition. Although HCO_3^- concentrations in soils may be highly variable because of fluctuations in CO_2 partial pressures, concentrations on the order of a few hundred microequivalents per liter might typically be expected in moderately acidic systems. As acidity increases, the HCO_3^- concentration drops off rapidly. At 3% CO_2 (about 100 times atmospheric CO_2, not unusual in soils), the HCO_3^- would be about 50 μeq/L at pH 5.0 and 5 μeq/L at pH 4.0. In general, we can conclude from Fig. 2.2 that unless soil CO_2 levels are more than 10 times atmospheric levels, significant leaching of bases in association with HCO_3^- probably will not occur if soil solution pH is <5.0. Even at 100 times atmospheric CO_2, HCO_3^- leaching would effectively cease below about pH 4.7.

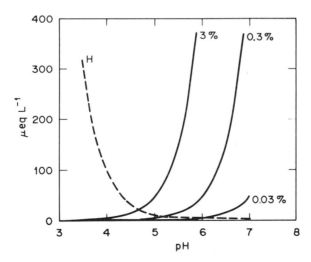

Figure 2.2. The activity (μeq/L) of the H^+ ion (broken line) and HCO_3^- ion (solid lines) as a function of pH. HCO_3^- activity is given for 0.03, 0.3, and 3% CO_2.

From these considerations, it appears that the practical lower limit for significant base loss or soil acidification resulting from carbonic acid leaching should be about pH 4.5. However, HCO_3^- is sometimes observed in solutions from soil horizons in which soil pH has been measured as substantially below 4.5. For instance, Johnson et al. (1983) found HCO_3^- levels of 100 to 300 μeq/L in solutions from Costa Rican and Tennessee soils having soil pH (water) of 3.8 to 4.5. The pH of these solutions was near 6, as would be expected when solutions containing HCO_3^- are removed from the soil and are allowed to equilibrate with atmospheric CO_2 levels (see Chap. 6). There seem to be two possible explanations for this apparent inconsistency. First, the CO_2 levels may be high in these tropical or warm temperate soils because of high decomposition rates. Second, observed soil pH values may be influenced by exchange acidity and do not necessarily accurately reflect soil solution pH. The overall implication of the CO_2–HCO_3^- equilibrium as illustrated in Fig. 2.2 is that soil acidification by leaching of base cations in association with HCO_3^- is self-limiting, and beyond a certain point, acidification caused by base cation leaching requires that other sources of anions be present.

In most of the natural systems with which we are concerned, the supply of mineral anions such as SO_4^{2-} or NO_3^- is very limited. In coarse-textured soils having thick, slowly decomposing, humus layers, organic acid anions assume an important role in soil leaching processes. Organic acids are responsible for the formation of Spodosols, which are found in boreal regions as well as in certain very sandy soils of the tropics and subtropics (Kononova 1966). These acids can produce low solution pH, provide counter-anions for cation leaching, and are responsible for the chelation and transport of iron and aluminum from surface (E) to subsurface (Bs) horizons during the process of podzolization. Like carbonic acid, organic acids tend to become protonated at low pH. However, because of the diversity of these acids, their behavior at low pH is considerably more complex than that of carbonic acid and is not well understood. Decreasing pH changes the structure of fulvic acid, causing it to aggregate and to lose some of its chelation ability, but may not necessarily cause its precipitation. Fulvic acid precipitation seems to be regulated primarily by the availability of iron and aluminum rather than pH (Peterson 1976). Unfortunately, we do not know enough about organic acids to formulate models of their behavior.

In contrast to the naturally acidified systems in which the supply of anions is likely to be limited by the protonation of the bicarbonate and organic acids, soils that are affected by sulfur deposition are supplied with an outside source of the SO_4^{2-} anion. In our opinion, this is a key difference between the acidification of natural and acid-affected systems. The implications of this difference must be carefully examined to understand the possible implications of long-term acid deposition impact.

Precipitation acidified to pH 4.0 by H_2SO_4 contains 100 μeq/L of SO_4^{2-} in excess of any basic cations that might be present. In most acid-affected systems, dry deposition (SO_x, etc.) adds to the total SO_4^{2-} burden because such forms are rapidly oxidized to SO_4^{2-} in the soil. A typical loading might be 1.25 m of pH 4.2 rainfall plus an equal amount of sulfur in dry deposition, for a total of 0.16 eq/m^2 of both H$^+$ and SO$_4^{2-}$ annually. Sulfate ion concentrations in soil solution would vary markedly through the season depending on the degree to which the soil solution is concentrated through evapotranspiration and other factors, but affected systems could typically expect 100 to 300 μeq/L (Seip 1980; Mollitor and Raynal 1982).

Increasing the strong acid anion concentration by 100 to 300 μeq/L will have a very significant effect on the anion status of most acid soils. The potential to leach base cations from a soil at solution pH 4.5, which would naturally have a low base status, would, under the influence of acid deposition, be the same as the natural bicarbonate leaching potential of soils with solution pH of 5.5 or greater (Fig. 2.2). Because the principle of electroneutrality requires an equivalent number of cations and anions in solution, we can also expect changes in the cation concentrations. Therefore, to predict the effect of acid deposition on the soil system, we must use our knowledge of soil processes to (1) evaluate the probable effects of this deposition on the concentration of the various ions in the soil solution and (2) project from this the changes that can be expected in the chemistry of the surface waters and of the soil itself. These issues will be addressed in the following chapters.

3. The Sulfur System

Inherent in the discussion in Chap. 2 is the assumption that sulfur deposition will directly increase the concentration of the SO_4^{2-} ion in solution. The response of the solution concentration to H_2SO_4 or other forms of sulfur deposition is complicated by biological effects and by the capacity of many soils to retain SO_4^{2-}. These processes can delay the onset of effects such as base removal, aluminum mobilization, and decreased alkalinity of the percolates for many years or even decades (Lee and Weber 1982). Failure to observe these effects in short- to medium-term experiments may delay, but by no means preclude, the likely long-term consequences in soils and surface waters.

3.1 The Sulfate Cycle

The sulfur system represented in the simplified sulfur flow diagram in Fig. 3.1 includes three inputs, one output, and four internal reservoirs. For an unaffected system, we can assume that over time the system has approached equilibrium between inputs and outputs. Under acid deposition stress, the equilibrium would be disturbed by the addition of H_2SO_4 and the deposition of SO_x and sulfates on the soils and plants (i.e., the inputs in Fig. 3.1). The output is SO_4^{2-} in the leachate, whereas the reservoirs include soil-solution sulfates, soil-adsorbed sulfate, sulfate in living plant material, and sulfate in litter and organic matter of the soil.

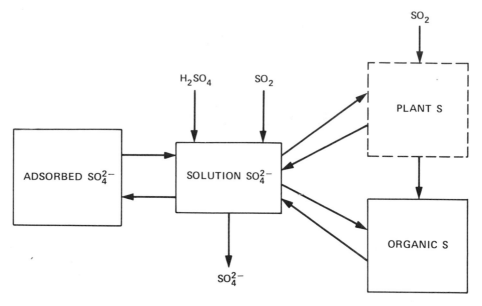

Figure 3.1. Simplified sulfur flow diagram for a forest ecosystem.

First let us consider the inputs. Sulfates in rainfall or in dry deposition either enter the soil solution directly or are deposited in the canopy and enter the soil as throughfall or stem flow during rainfall. We can assume that the SO_2 adsorbed by the soil will be rapidly oxidized to SO_3 and, thus, be equivalent to an input of H_2SO_4 in precipitation. The SO_2 absorbed by plants will either be reduced to sulfhydryl groups in the plant tissue or oxidized to SO_4^{2-}. If sulfate remains within the intercellular spaces within the leaves, it will be subject to leaching by subsequent rainfall. On death and decomposition of the plant material, the sulfhydryl groups will be oxidized to SO_4^{2-}, so that, effectively, all input becomes SO_4^{2-} in the soil. Although transport of part of the sulfur may be subject to various delays because of biological incorporation, it is probably safe to assume that most reaches the soil solution within the same annual cycle as it was deposited.

The reservoirs in the system are the SO_4^{2-} adsorption sites in the soil, the living plant tissue, soil organic matter and litter, and a relatively small soil-solution reservoir. Given sufficient time at any level of deposition, we can expect that the reservoirs will come to equilibrium with the new level of input, so that the flux downward from the root zone will be approximately equal to the input. Therefore, in the long term, the input of H_2SO_4 and SO_2 can be used to estimate the acidification potential. Unfortunately, the time required to approach an input–output equilibrium will be highly dependent on soil properties, particularly the SO_4^{2-} retention capacity, and could be anywhere from a few weeks or months to several decades. For time scales less than that required to approach an equilibrium, the capacity of these reservoirs may very significantly affect the flux. An understanding of these effects is particularly important to interpretation of relatively short-term experimental results.

3.2 Sulfate Adsorption

When a soil is subjected to acid deposition, the concentration of SO_4^{2-} in solution increases. However, SO_4^{2-} adsorption is a concentration-dependent process (i.e., the capacity to adsorb SO_4^{2-} increases with solution SO_4^{2-} concentration; Chao et al. 1962a,b). The functional relationship between solution and adsorbed SO_4^{2-} is known as the sulfate adsorption isotherm. A schematic example of this relationship is shown in Fig. 3.2 (Johnson and Cole 1980). For any given solution concentration, SO_4^{2-} will adsorb on the soil sesquioxide surfaces until the corresponding soil-adsorbed-sulfate value is reached on the isotherm. When that point is reached, the soil should be in steady state in which outputs equal inputs. In the case in which SO_4^{2-} inputs increase, concentrations increase, thereby activating "new" adsorption sites and causing a net SO_4^{2-} retention in the soil. With continued inputs, a new steady-state condition will eventually be reached.

Several mathematical formulations have been used to describe this concen-

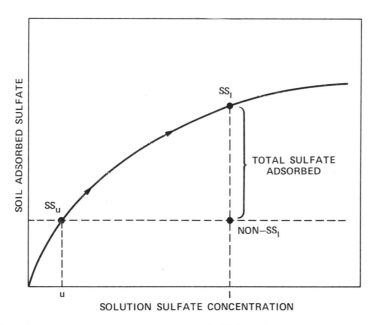

Figure 3.2. Schematic representation of a soil SO_4^{2-} adsorption isotherm. U, undisturbed soil conditions, I, soil conditions following sulfate impact; SS and NON-SS refer to steady-state and non-steady-state conditions, respectively (From Johnson D. W. and Cole D. W. *Environ. Int.* 3:79–90, 1980.)

tration-dependent relationship, the most common being that known as the Langmuir isotherm, one form of which is

$$S_a = \frac{K_m S_s}{K_s + S_s},$$
(3-1)

where S_a and S_s are the adsorbed (mol/kg) and solution (mol/L) SO_4^{2-} respectively, and K_m and K_s are constants. K_m represents the mol/kg adsorbed at infinite solution concentration, whereas K_s is the concentration at which S_a is equal to one-half K_m.

If the SO_4^{2-} adsorption process is completely reversible, the adsorption and desorption isotherms will be identical. Unfortunately, we cannot assume that this will be the case because some SO_4^{2-} may be irreversibly adsorbed, resulting in a desorption isotherm lying above the adsorption isotherm. A schematic representation of completely reversible (line a), partially reversible (line b), and irreversible (line c) isotherms is shown in Fig. 3.3.

The concentration-dependent relationship between solution and adsorbed SO_4^{2-} will result in a "front" moving downward through a SO_4^{2-} adsorbing soil when a new, higher level of SO_4^{2-} input is introduced and continually maintained. Soil above (or behind) the front will have a new higher level of adsorbed SO_4^{2-} in response to the increased solution levels. Soil solution samples taken above the front might indicate movement of cations and SO_4^{2-} whereas samples from a lower depth would show essentially no movement of cations and SO_4^{2-}.

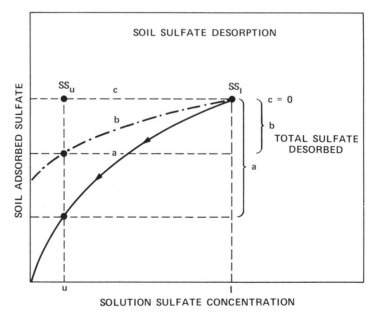

Figure 3.3. Schematic representation of a (a) reversible, (b) partially reversible, and (c) irreversible SO_4^{2-} adsorption isotherms. U, undisturbed (or final) soil conditions; I, soil conditions under SO_4^{2-} impact; SS refers to steady-state conditions.

Thus, a SO_4^{2-} adsorbing soil delays cation leaching effects of dilute H_2SO_4 inputs until the adsorbing capacity, which is dependent on input concentration, is satisfied down through the soil zones of interest. From the standpoint of base removal, the time prior to increased SO_4^{2-} concentrations below the root zone is essentially a grace period because the basic cations remain available for plant uptake and recycling in the system. The SO_4^{2-} adsorption causing this delay can also be considered a neutralization of applied acid because soil adsorption of SO_4^{2-} from applied H_2SO_4 solutions is often accompanied by a rise in pH, apparently resulting from the replacement of $-OH^-$ groups on the mineral surface by SO_4^{2-} (Chao et al. 1965). At low pH, no pH change may be observed, probably because $-OH_2^+$ groups are replaced (Rajan 1979).

As a consequence of the SO_4^{2-} front phenomenon, time traces of soil solution SO_4^{2-}concentrations in systems subjected to H_2SO_4 deposition may often assume a sigmoidal form (Fig. 3.4). After passage of the SO_4^{2-} front, the higher solution concentration will be maintained in approximate equilibrium with the affected level of input. After equilibrium has been attained, soil solution SO_4^{2-} concentration will normally be higher than that found in the rainfall because of concentration by evapotranspiration. The lag time before elevated SO_4^{2-} concentrations occur in the soil solution after an increase in rainfall SO_4^{2-} will, thus, depend on the SO_4^{2-} adsorption properties of the soil and the depth at which the solution is sampled.

Galloway et al. (1983) describe a seven-stage scenario very similar to that

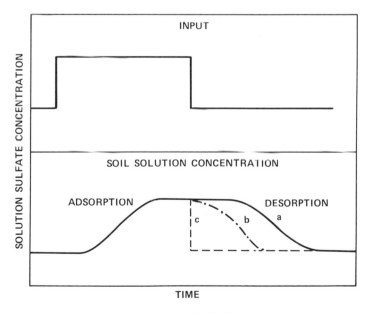

Figure 3.4. Schematic of effect of changes in SO_4^{2-} input concentration over time on soil solution SO_4^{2-} concentration, assuming (a) completely reversible SO_4^{2-} adsorption, (b) partial reversibility, and (c) irreversible adsorption.

discussed above, except for changes in soil base saturation in their model. Stage 1 is preacidification; stage 2 is a SO_4^{2-} adsorption phase; stage 3 is SO_4^{2-} breakthrough accompanied at first primarily by accelerated base cation leaching; then as base saturation (BS) is reduced, an increase in solution acidity occurs. Stage 4 is a steady-state, acidified condition in which "% BS values are near zero;" stage 5 is indicated by a reduction in SO_4^{2-} input and desorption of SO_4^{2-} from soils; stage 6 is signified by recovery (or increase) in % BS; and stage 7 is a period of stable recovery. We have added some complexities to Galloway's model (discussed above and in Chap. 7) and are unsure of the predicted changes in % BS (because weathering rates are unknown and may change in response to changes in leaching), but basically we concur with and support the fundamental concepts therein.

If the new equilibrium concentration can be estimated, it is possible to estimate the time to passage of the SO_4^{2-} front for a given depth and level of SO_4^{2-} deposition by calculating the difference between the amount of adsorbed SO_4^{2-} before deposition and the amount adsorbed after equilibrium is achieved. This calculation uses Eq. (3-1) and requires the empirical laboratory determination of the constants K_m and K_s. When a SO_4^{2-} adsorbing Bohannon silt loam soil was subjected to artificial H_2SO_4 rainfall, the SO_4^{2-} front arrival at 20 cm occurred in about 18 to 24 months (Lee and Weber 1982). In these experiments, no elevation in SO_4^{2-} concentration was found at the 1-m depth over the 42-month duration of the experiment. Calculations by the senior author of the time to passage of the front based on laboratory-determined constants for this soil give estimates of about 2 years for the 20-cm depth and roughly a decade for the 1-m depth of the lysimeters.

As has been previously pointed out (Johnson and Cole 1980), the time required to reach equilibrium is dependent on the slope of the isotherm. It should be further emphasized that, as a consequence, the time required for passage of the SO_4^{2-} front to any particular depth is likely to be relatively insensitive to the strength of H_2SO_4 affecting the system. The time will be affected by the acid strength only if the slope of the isotherm changes substantially over the range of concentration differences occurring in response to acid deposition, and abrupt slope changes are seldom observed. This point is important for interpreting experiments in which plots or lysimeters are artificially exposed to high loadings to accelerate base removal and acidification. The fact that several years may be required before elevated concentrations of SO_4^{2-} are observed in soil solutions in a highly loaded system does not mean that a much longer time will be required for elevated concentrations to occur at more realistic input levels.

The question of reversibility of the SO_4^{2-} adsorption process is crucial to our ability to predict the long-term effects of acid deposition, particularly the response to a decrease in input (Fig. 3.4). If SO_4^{2-} is irreversibly adsorbed, decreased levels of input will be reflected rapidly in decreased soil-solution SO_4^{2-} levels (line c) and a cessation of any accelerated cation leaching that may have resulted from acid deposition. On the other hand, to the extent that reversibly adsorbed SO_4^{2-} is present (lines a and b), the reduction of solution concentration and cation export will lag behind any reduction in acid deposition. In this case, SO_4^{2-} will desorb from the soil to a point on the isotherm at which the solution SO_4^{2-} concentration is in equilibrium with the new input level. During desorption, output exceeds input and the SO_4^{2-} and cations previously retained during adsorption are leached from the soil.

The reversibility of SO_4^{2-} adsorption varies with soil properties and the desorbing solution used. In some cases, water extraction recovers all adsorbed SO_4^{2-}, whereas in other cases, full recovery is achieved only with extractants such as phosphate or acetate that have a greater affinity for the adsorption sites than SO_4^{2-} (Harward and Reisenauer 1966). Even phosphate does not always replace all adsorbed SO_4^{2-}, as shown by Bornemisza and Llanos (1967) for highly weathered tropical soils. Currently, it would seem that our limited knowledge of the extent to which this process will prove to be reversible is a serious obstacle to the prediction of the lags that are likely to be associated with recovery of the system if deposition is reduced.

The above discussion assumes that the major storage reservoir for inorganic SO_4^{2-} in the soil is SO_4^{2-} adsorption and that this adsorption can be quantitatively described by the use of a classic adsorption isotherm such as the Langmuir. Some workers (e.g., Adams and Rawajfih 1977; Nilsson and Bergkvist 1983; Nordstrom 1982; Prenzel 1983) have suggested that the precipitation of certain aluminum-SO_4^{2-} minerals may be involved. Although from an empirical or operational standpoint it may be difficult to establish which of these mechanisms is responsible for SO_4^{2-} accumulation in a particular situation, the point is nonetheless important. A more detailed discussion of these reactions is given in Chap. 5.

3.3 The Biotic Component

Further lags in equilibrium will occur as a result of flows into and out of the biomass component. Calculation of this lag is perhaps more difficult, but reasonable estimates can be made. Biological sulfur requirements for forests are modest, generally <5 kg $\cdot$ ha^{-1} $\cdot$ year^{-1} for the net vegetative increment. (Vegetative uptake in forest systems may range from perhaps 5 to 25 kg $\cdot$ ha^{-1} $\cdot$ year^{-1}, but most of this is recycled and, therefore, does not represent a net retention.) Atmospheric inputs of sulfur in polluted regions are generally well in excess of 10 kg $\cdot$ ha^{-1} $\cdot$ year^{-1} and can go as high as 80 kg $\cdot$ ha^{-1} $\cdot$ year^{-1}. These inputs tend to exceed not only the forest ecosystems sulfur requirements but also the ability of forest vegetation to accumulate sulfur (Johnson et al. 1985). There is some increase in sulfate-sulfur content of forests in response to elevated atmospheric inputs, but apparently this capacity is easily reached, and in most affected systems, accumulation in the forest biomass represents a relatively minor part of the capacity of the system to accumulate sulfur.

The sulfur contained in soil organic matter represents the largest pool of sulfur in most forest ecosystems, and most organic sulfur occurs in either ester SO_4^{2-} or carbon-bonded sulfur forms (Bettany et al. 1979). Soil organic sulfur pools range from <100 kg/ha in sulfur-limited forest ecosystems in Australia (Turner and Lambert 1980) to over 4000 kg/ha in heavily affected ecosystems such as the Walker Branch Watershed in Tennessee (Johnson et al. 1982a). Recently, Fitzgerald et al. (1982) found SO_4^{2-} conversion to ester- and carbon-bonded sulfur forms in a deciduous forest soil from Coweeta Watershed in North Carolina, and their assertion that this process must be considered in future attempts to address ecosystem sulfur retention is well founded. Although soil organic matter usually contains most of the sulfur found in soils, adsorbed SO_4^{2-} assumes increasing importance in soils heavily loaded by acid deposition and may actually exceed organic sulfur in soils receiving large atmospheric inputs. For instance, Meiwes and Khanna (1981) report that SO_4^{2-} constitutes 66% (950 kg/ha) of total soil sulfur (1450 kg/ha) in a spruce stand at Solling, West Germany.

There is no doubt that SO_4^{2-} can be immobilized by incorporation in organic matter. Empirical evidence, however, suggests that SO_4^{2-} adsorption is a more important accumulation mechanism than biological immobilization. Forest ecosystems on the SO_4^{2-} adsorbing Ultisols (red-yellow podzolic soils) of the southeastern United States generally show a net accumulation of sulfur, whereas more northerly ecosystems having soils much richer in organic matter but poor sulfate adsorption capacities generally show a balance (i.e., sulfate inputs and outputs are nearly equal). This effect can be clearly seen in Table 3.1, which shows the ratio of sulfate outputs in streamwater to sulfate in bulk

Table 3.1. Sulfate and Water Outflow/Inflow Flux Ratios and Sulfate Trend Slopes for 16 Headwaters Watersheds in the Eastern United States

	Flux Ratio Outflow/Inflow		SO_4^{2-} Trend Slope [μmol (eq) L^{-1} year^{-1}]	Significance Level
	H_2O	SO_4^{2-}		
Gilead, Maine	0.71	1.42	−1.7	0.000
Shandaken, New York	0.72	1.79	−2.4	0.000
Renovo, Pennsylvania	0.55	1.16	−0.9	0.088
Andersonville, Virginia	0.26	0.29	−1.0	0.040
Bishopville, South Carolina	0.31	0.37	2.9	0.001
New Ellenton, South Carolina	0.39	0.15	1.0	0.007
Juliette, Georgia	0.23	0.36	0.6	0.696
Sopchoppy, Florida	0.37	0.72	7.8	0.000
Grayson, Alabama	0.45	0.80	1.1	0.016
Janice, Mississippi	0.41	0.36	2.3	0.002
Calatoochee, North Carolina	0.62	0.21	1.0	0.031
Flatwoods, Tennessee	0.44	0.64	−0.7	0.048
McGaw, Ohio	0.31	2.10	11.6	0.000
Dillsboro, Indiana	0.34	5.41	−4.3	0.239
Windigo, Michigan	0.58	1.75	−1.3	0.308
Fence, Wisconsin	0.40	0.40	−2.1	0.034

We thank Dr. Robert Rosenthal for suggesting this analysis.
Data from Smith and Alexander (1983).

precipitation inputs for 16 headwater watersheds in the United States east of the Mississippi River, as calculated from data reported by Smith and Alexander (1983). One station (Lebanon State Forest, New Jersey) was deleted because the water flux ratio clearly showed that a high percentage of input water was leaving the system by some pathway other than the measured channel. Sulfate flux ratios for the three stations in the northeastern United States are >1.0, which indicates that either stored SO_4^{2-} is being discharged from the system or that significant sulfur is entering the watersheds, other than that measured in precipitation, probably as dry deposition. The nine watersheds reported in the southeastern United States all show flux ratios <1.0, indicating that these watersheds are accumulating SO_4^{2-}. No consistent pattern is apparent for the four watersheds reported from the upper Midwest. The results for the northeastern and southeastern regions are consistent with the hypothesis of Johnson and Todd (1983) that Spodosols, commonly found in the Northeast, are generally inefficient SO_4^{2-} adsorbers, whereas SO_4^{2-} tends to be strongly adsorbed in the Ultisols commonly encountered in the southern United States. It is also noteworthy that lysimeter budgets for ecosystems accumulating SO_4^{2-} usually show a net retention (input versus output) only in the SO_4^{2-}-adsorbing B horizons and not in the organic-matter-rich A horizons (Johnson et al. 1982a; Singh et al. 1980; Stednick 1982).

3.4 Summary

We have described a protocol for estimating SO_4^{2-} flux through the system and the time lags that would be associated with this flux. The need for SO_4^{2-} isotherms in this protocol imposes a serious limitation because of the limited availability of input data. This limitation is accentuated by the variation of SO_4^{2-}-adsorbing properties with depth, particularly in Ultisols and Spodosols (Singh et al. 1980; Johnson and Todd 1983). There seems to be no reasonable alternative to the experimental determination of adsorption isotherms, at least for the most common soil series. This could be supplemented by a concerted effort to relate SO_4^{2-} adsorption properties, including reversibility, to characteristics that are routinely available from standard soil surveys.

4. The Nitrogen System

Although both nitrogen and sulfur are components of protein, and, thus, are essential elements for plant nutrition, there are some fundamental differences in the cycles of these two elements in terrestrial ecosystems that must be considered in our analysis of soil acidification. In contrast to the sulfur system in which SO_4^{2-} adsorption can be an important factor in the accumulation of sulfur in the ecosystem, there is little tendency for inorganic nitrogen to accumulate in most forest soils. The most oxidized form of nitrogen, the nitrate (NO_3^-) anion, is only weakly adsorbed on soil surfaces and is, thus, readily leached. The ammonium ion (NH_4^+) can be retained on cation exchange sites. However, the quantities of exchangeable NH_4^+ on exchange sites in forest soils is usually small because (1) high biological demand by plants and microorganisms in nitrogen-limited ecosystems (which are very common in forests of North America and Scandinavia) causes NH_4^+ to be rapidly taken up and (2) any NH_4^+ that is not incorporated into biological tissue is likely to be oxidized by microorganisms to nitrate. Although this oxidation is commonly thought to be inhibited under acid conditions, recent reports of high nitrification rates in very acid forest soils under heavy loadings of acid deposition (van Breemen et al. 1982; van Breemen and Jordens 1983) testify to the capacity of the system to oxidize NH_4^+. Because large pools of inorganic nitrogen are seldom, if ever, found in forest ecosystems, we can safely generalize that, for nitrogen, biological rather than chemical mechanisms account for ecosystem accumulation. We can further assume that, in humid regions, most nitrogen deposited in excess of biological demands by plants and heterotrophic microorganisms will ultimately be lost as NO_3^-. [In some (i.e., in poorly drained) soils excess nitrogen may be lost by denitrification as well as by NO_3^- leaching.]

4.1 Acid–Base Relationships of the Nitrogen Cycle

The acid–base relationships of the nitrogen cycle are complex but are reasonably well known. The nitrogen cycle and its associated nitrogen fluxes are depicted schematically in Fig. 4.1. Beginning with soil organic nitrogen, we will follow the H^+ production and consumption for each step in the cycle, as well as the net cumulative H^+ production, and discuss the effects of atmospheric nitrogen inputs on these processes.

The formation of ammonia (NH_3) during the decomposition of proteins does not involve the production or consumption of H^+ ions. However, the second step of this "ammonification" process is the protonation of NH_3 to form NH_4^+. This process can be thought of as either the reaction of water with NH_3 to form NH_4OH (i.e., the production of an OH^- ion) or the consumption of an H^+ ion via the direct protonization of NH_3 to form NH_4^+; in aqueous solutions, the processes are equivalent. Thus, there is a net cumulative H^+ production (deficit) of -1 at this point. If the NH_4^+ is taken up by a plant or a microorganism, an H^+ is released in the uptake process, so that the net cumulative H^+ production is zero. If NH_4^+ is oxidized to NO_3^-, 2 H^+ ions are released for each NO_3^- formed, making the net cumulative H^+ production $+1$ (i.e., 1 mole of HNO_3 is produced for each mole of NH_3 that is transformed to NO_3^-).

The fate of any NO_3^- formed is of crucial importance in determining the effect of the nitrogen cycle on soil acidification. If NO_3^- is taken up by plants, OH^- is released, neutralizing H^+ and the net cumulative H^+ production is again zero. However, if NO_3^- leaches from the system the net cumulative H^+ pro-

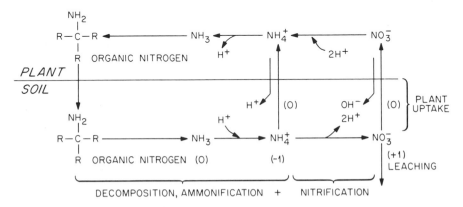

Figure 4.1. Acid–base relationships of the nitrogen cycle. Numbers in parentheses refer to the net production ($+$) or consumption ($-$) of H^+ ions in the soil system starting with soil organic nitrogen.

duction remains at $+1$. Following our conceptual model of Chap. 3, H^+ will react with whatever mineral phase is controlling the solubility of aluminum in the soil to produce an ionic aluminum species that will, except in the most acid soils, replace base cations on the ion-exchange complex. Nitrate salts will be leached and the soil acidified. The distribution of cations in solution will be determined by the relationships discussed in Chap. 5, such that in very acid systems, a significant fraction of the solution cations would be aluminum species rather than base cations.

Within the plant, when NO_3^- is taken up, an OH^- ion is given off, which is equivalent to the production of 1 H^+. Because the net conversion of NO_3^- to the neutral NH_3 (or $R–NH_2$) requires a net input of one H^+, the cycle is balanced. If nitrogen is taken up as NH_4^+, the H^+ given off in the uptake process is balanced by the release of an H^+ in the conversion of NH_4^+ to the neutral NH_3.

In summary, the net H^+ production of the natural nitrogen cycle is zero if (1) no nitrate leaching occurs and (2) no external inputs of nitrogen (e.g., atmospheric) occur. Although we have not discussed the situation within the plant in detail, the cycle is, in fact, balanced both within the plant and within the soil. Even though the decomposition/uptake cycle is balanced with respect to H^+ production, natural nitrogen cycles can be quite acidifying, even in pristine areas, if the rates of NO_3^- production exceed biological uptake so that leaching of NO_3^- occurs. This situation is known to occur in nitrogen-fixing forests. For example, van Miegroet and Cole (1984) have clearly documented high rates of nitrification and soil acidification caused by leaching of excess NO_3^- in red alder (*Alnus rubra*) in western Washington. Thus, not all natural nitrogen cycles are balanced with regard to acid–base relationships, although, by and large, this should be the case in nitrogen-deficient forests where leaching of significant quantities of NO_3^- is very rare.

4.2 Acid–Base Relationships of Nitrogen Inputs

On a broad scale, the balance between nitrogen supply and the biological demand for nitrogen generally controls NO_3^- leaching, and, thus, controls the consequent effects of nitrogen deposition on soil acidification via base cation leaching. However, external nitrogen inputs, whether by nitrogen fixation, fertilization, or acid deposition, can fundamentally affect the acid–base relationships of the nitrogen cycle. The form in which nitrogen is deposited (i.e., organic, NO_3^-, or NH_4^+), as well as the associated anion(s) or cation(s), may also have a direct bearing on the net of H^+ production or consumption in the soil. Therefore, we will consider in detail the effect of inputs in the form of HNO_3, $(NH_4)_2SO_4$, and NH_4NO_3 (the forms of nitrogen inputs most often observed as components of acid deposition) and of the subsequent transformations that occur in the soil. Urea (or ammonium bicarbonate) and ammonium phosphate forms are important in regard to nitrogen fertilization but will not be discussed here.

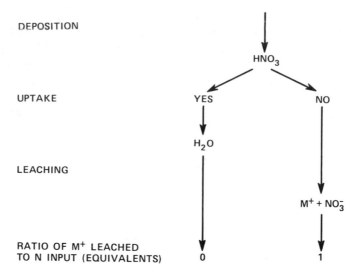

Figure 4.2. Acidification effects associated with nitric acid (HNO_3) inputs.

Considering first HNO_3 input (Fig. 4.2) an equivalent amount of base cation leaching will occur in response to HNO_3 input only if NO_3^- remains mobile. In the event that NO_3^- is taken up by plants or microorganisms, OH^- is released in the uptake process (Fig. 4.1), neutralizing the incoming H^+ to form water, and, thus, no net cation leaching occurs.

Now, let us consider the effects of the input of $(NH_4)_2SO_4$ (Fig. 4.3), a form

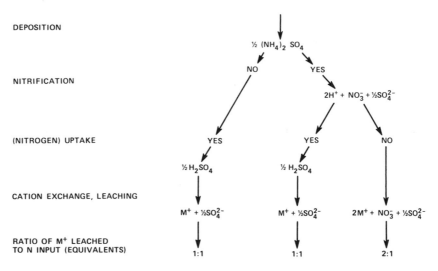

Figure 4.3. Acidification effects associated with ammonium sulfate [$(NH_4)_2\,SO_4$] inputs.

that could be expected where both SO_2 and NH_3 emissions occur simultaneously (e.g., van Breemen et al. 1982). In this case, the NH_4^+ in incoming $(NH_4)_2SO_4$ can either be taken up by plants or nitrified. If it is taken up, H^+ is released and the effect is equivalent to the introduction of 1 equivalent of H_2SO_4 directly into the soil per equivalent of $(NH_4)_2SO_4$ input. If NH_4^+ is nitrified and then taken up, the net effect is exactly the same because 2 H^+ equivalents are released during nitrification for each NH_4^+ nitrified (Fig. 4.1), and 1 H^+ is neutralized by the OH^- released during uptake, giving a net H^+ production of 1 equivalent. If NO_3^- is not taken up, a mixture of HNO_3 and H_2SO_4 remains, both of which provide mobile anions (assuming SO_4^{2-} mobility) for leaching of cations. The net effect is to leach 2 equivalents of cations from the soil per equivalent of NH_4^+ input or 4 equivalents/mol of $(NH_4)_2SO_4$. In summary, it is important to understand that the minimum potential for acidification per mole of $(NH_4)_2SO_4$ input is equal to the acidification potential of 1 mol of H_2SO_4, whereas the maximum acidification potential is equal to that of 2 mol of H_2SO_4 depending on whether the nitrogen is taken up or nitrified to NO_3^- and subsequently leached from the system.

Finally, let us consider the effects of NH_4NO_3 inputs. Although a number of possibilities exist because both NH_4^+–N and NO_3^-–N are present in equimolar amounts in this salt, the simplest model is probably that shown in Fig. 4.4. If nitrification does not take place and NH_4^+ is taken up by plants or microorganisms, one H^+ ion is released during the uptake process. This release, when combined with the NO_3^- ion in the input, can be thought of as an input of 1 mol

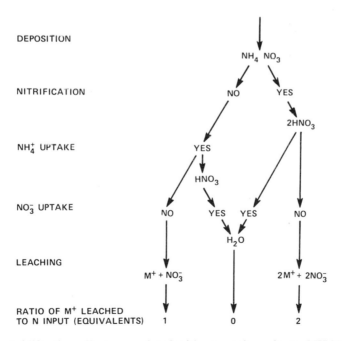

Figure 4.4. Acidification effects associated with ammonium nitrate (NH_4NO_3) inputs.

of HNO_3 for each mol NH_4NO_3 input. On the other hand, if the NH_4^+ is oxidized to NO_3^- in the soil, 2 H^+ ions are released in the oxidation process (Fig. 4.1), so that the total effect is the same as an input of 2 mol of HNO_3. If NO_3^- is taken up from HNO_3, there will be no soil acidification from the input, whereas if the NO_3^- is leached along with a basic cation, acidification will occur. Therefore, the extent to which acidification resulting from leaching of bases occurs as a result of NH_4NO_3 inputs depends entirely on the extent to which the nitrogen is lost through NO_3^- leaching.

4.3 Ecosystem Effects

Some rather interesting questions arise concerning the above processes when one considers possible constraints imposed on the system by inhibition of the nitrification process and/or selective use of either NH_4^+ or NO_3^- by some plant species. For instance, as mentioned above, the generally accepted view that acid soils inhibit the oxidation of NH_4^+ to NO_3^-, when coupled with the also generally accepted view that many tree species prefer or even require the NH_4^+ form of nitrogen (van den Driessche 1971), gives rise to a concept that many forest ecosystems on acid soils actually operate on a nitrogen cycle dependent on only the NH_4^+ form of nitrogen. If a system operating under such constraints were exposed to a mixture of NH_4^+ and NO_3^- inputs from acid deposition, the NO_3^- would simply pass through to the drainage water, taking with it a complement of base cations, whereas the NH_4^+ would either be taken up by the plants and microorganisms or, in a system that is not nitrogen-deficient, would accumulate as exchangeable ammonium. To some degree, this scenario may exist in affected systems. However, in our view, these constraints are by no means absolute. Thus, very low $NO_3^- - N$ levels commonly found in leachates from nitrogen-deficient forests (even under moderate atmospheric nitrogen loadings) suggest that most systems probably have at least some capacity to utilize NO_3^-. Conversely, the high NO_3^- outputs observed from very acid soils under heavy loadings of $(NH_4)_2SO_4$ (van Breemen et al. 1982; van Breemen and Jordens 1983) would suggest that these soils do, in fact, have substantial capacity to nitrify.

In the preceding discussion, we have presented a rather detailed view of the acid–base relationships of certain relevant soil processes. In so doing, we have invoked the concept of H^+ and OH^- ion releases by organisms during the uptake of cations and anions, respectively. The reader could easily gain the impression that by using this conceptual model we are regarding the organisms as some inexhaustable source or sink of H^+ and/or OH^-. Certainly, this is not the case. The nitrogen cycle as depicted in Fig. 4.1 is balanced in regard to H^+, both within the soil system and within the plant. Thus, the synthesis of $R-NH_2$ compounds, starting with either NH_4^+ or NO_3^-, should produce the required number of H^+ (or OH^-) for exchange.

It is apparent from the preceding discussion and Figs. 4.1 through 4.4 that the reactions of N in the soil are complex. There are many pathways, each having its own set of acid–base relationships. If one follows these pathways correctly, a balance sheet of the H^+ ion production and consumption can be drawn that will reflect accurately the acidification potential of the nitrogen system. However, an analysis of this type is subject to errors at each step, and, thus, the cumulative error may be substantial.

A simpler, and in our view generally more useful, model is available in the mobile anion concept. If one examines the above nitrogen pathways closely, it is apparent that the net acidification potential is expressed as the number of mobile anions available for leaching in association with base cations. If nitrogen inputs are in the form of NH_4^+ salts associated with mobile anions such as Cl^- or SO_4^{2-} (to the extent SO_4^{2-} is mobile in the soil), it is inevitable that leaching of base cations will occur in association with these anions, except in very acid soils where a substantial fraction of the cations leached will be aluminum species and/or H^+ (Chap. 5). The potential of the nitrogen itself to acidify the soil through the base leaching mechanism depends entirely on the fraction of the nitrogen that is leached in the form of NO_3^- and is virtually independent of the form in which the nitrogen enters the soil. In general, nitrogen-rich systems will be subject to NO_3^- leaching, and this leaching will be accelerated by additional atmospheric nitrogen inputs of any form, except perhaps very recalcitrant organic nitrogen. Nitrogen-limited systems will not be subject to substantial acidification by accelerated NO_3^- leaching, at least from moderate outside inputs. Heavy inputs, of course, would result in a system that is no longer nitrogen-limited.

Nitrogen inputs may cause acidification of nitrogen-limited systems through an entirely different mechanism, even though all additional nitrogen is taken up and no leaching occurs. This would occur as a result of increased growth and consequent increases in demands for Ca^{2+}, K^+, and Mg^{2+} from the soil. It is very difficult to generalize about whether or not such accelerated base uptake will occur in a given system because the relationship between growth and base cation demand is not necessarily linear. For instance, Miller et al. (1979) found that nitrogen fertilization caused increased growth and uptake of all nutrients except calcium in nitrogen-deficient Corsican pine (*Pinus nigra*) stands in Scotland. They interpreted this as an indication of luxury consumption of Ca by nitrogen-deficient pines, such that increased growth resulted merely in a dilution of tissue calcium. In any event, we do not wish to imply that acidification by increased growth from nitrogen is likely to negate the potential benefits of enhancing the nitrogen status of nitrogen-deficient sites. There is little doubt that atmospheric nitrogen inputs could contribute to increased growth in forests of North America and Scandinavia. As long as atmospheric nitrogen inputs generally fall well short of tree nitrogen demands, substantial increases could probably be accommodated without suffering the effects of cation leaching resulting from excess nitrogen supplies. There is evidence, however, that nitrogen inputs in excess of 20 kg $\cdot$ ha^{-1} $\cdot$ year^{-1} in parts of central Europe and southern Scandinavia now exceed forest requirements (van Breemen et al. 1982; Persson 1982; Höfken 1983; van Breemen and Jordens 1983) and may be reaching a state of nitrogen saturation.

5. Soil-Solution Interactions

To understand what changes in the soil may be brought about by a change in the level of atmospheric acid deposition, it is useful to develop at least a conceptual model of the major system components and the processes by which they interact. The interaction between the solid and solution phases in the soil is critical in determining the nature of these changes. Although the system in total is extremely complex, a relatively simple subset of the processes involved is sufficient for our purpose. In this chapter, we describe the major processes that control the composition of the soil solution and how this composition is modified by acid impact. Our approach is to consider the anions in solution and the processes by which their concentration is controlled. The principle of electro-neutrality requires the equivalence of positive and negative charges, so that if the anion concentrations are known, the total charge of the cations in solution is also known. We can then proceed to consider the processes by which the distribution of the various cations in solution is determined.

5.1 Role of Anions

The major anions of concern are HCO_3^-, SO_4^{2-}, NO_3^-, Cl^-, and organic anions. We shall postpone our discussion of organic anions and consider only the mineral anions here. The processes by which the concentration of SO_4^{2-} are controlled have been described in Chap. 3. Here, we shall simply reiterate that having reasonable knowledge of the SO_4^{2-} adsorption properties of the soil and the sulfur-cycling characteristics of the ecosystem one needs to know only the hydrological characteristics of the system to estimate the time response of SO_4^{2-} concentrations to changes in deposition. Although the lag times associated with various systems may be highly variable, most forested ecosystems subject to acid deposition contain sulfur well in excess of biotic requirements. Therefore, because of the additional input of dry deposition and concentration by evapotranspiration, the equilibrium soil-solution concentration is greater than that of the precipitation.

In continental systems, Cl^- concentrations are generally low or at least unaffected by anthropogenic inputs and can be estimated by assuming either a background concentration, or preferably, a constant flux so that the concentration varies inversely with evapotranspiration. In systems of heavy marine influence, the Cl^- inputs can be substantial, but over time these systems tend to be very nearly in balance (i.e., the output flux is equal to to the input), so that Cl^- concentrations of soil solutions and percolates can be reasonably estimated.

Nitrate is a common component of acid deposition, as is NH_4^+, which may be oxidized to NO_3^- in the soil. Although the nitrogen system is discussed separately in Chap. 4, recall that, in nitrogen-deficient ecosystems, biological uptake processes generally maintain NO_3^- at very low levels, so that the concentration of NO_3^- is not a major factor in determining the total anion strength in the soil solutions. In systems having heavy nitrogen inputs, whether from nitrogen fixation, fertilization, or acid deposition, the NO_3^- anion can contribute significantly to the total solution strength. In some cases, it may even be the dominant anion.

As shown by Eq. (2-1) (Chap. 2), the HCO_3^- ion concentration is regulated by the CO_2 partial pressure of the system and the H^+ ion concentration (pH) of the solution. The CO_2 partial pressure is controlled by the level of biological activity in the soil and the rate of diffusion out of the system (i.e., factors external to the soil solution). However, the H^+ concentration in solution is also affected by the CO_2–HCO_3^- equilibrium. This creates an interdependency as the total cation concentration is fixed by the total number of anions in solution. The concentration of one of these anions (i.e., HCO_3^-) is determined in part by the concentration of the H^+ cation, which in turn is determined in part by the CO_2–HCO_3^- equilibrium. Thus, HCO_3^- concentrations can be quantitatively estimated only by simultaneously solving Eq. (2-1) and the other equations that control cation concentration. Qualitatively, however, the effect of acid deposition will be to reduce HCO_3^-. This reduction results not so much from the acidity of the depositing solution as from the increased acidity resulting from the interaction of increased SO_4^{2-} with the cation-exchange system in the soil.

To summarize, the overall effect of H_2SO_4 acid deposition on anion concentration will be to increase the SO_4^{2-} (typically in the range of 100 to 300 μeq/L), whereas usually little change in the Cl^- and NO_3^- levels in solution will occur. The HCO_3^- concentration will be reduced, and this reduction may have highly significant effects, as will be described below.

5.2 Ion Equilibria Model

The base cations of concern are Ca^{2+}, Mg^{2+}, Na^+, and K^+. The major pools of these elements include that incorporated in living and dead organic materials, ions adsorbed on the negatively charged exchange complex, and that contained in the structure of soil minerals. For short-term soil-solution equilibria, we are concerned mainly with the exchangeable fraction. Whereas K^+ is of particular concern for plant nutrition, the divalent species Ca^{2+} and Mg^{2+} are usually dominant in terms of total charge. These two ions are chemically similar, so for the sake of convenience, we shall simply use Ca^{2+} as an approximation of $Ca^{2+} + Mg^{2+}$. The major nonbasic cations, at least in well-aerated soils, are H^+ and the ionic aluminum species, Al^{3+}, $Al(OH)^{2+}$, and $Al(OH)_2^+$. By considering only the anions SO_4^{2-}, HCO_3^-, OH^-, and Cl^- (or $Cl^- + NO_3^-$) and the cations H^+, Ca^{2+}, Al^{3+}, $Al(OH)^{2+}$, and $Al(OH)_2^+$, it is relatively easy to construct chemical equilibrium models that will demonstrate the effect of increased SO_4^{2-} concentration on the concentration of other ions in the solution.

For our purpose, a description of only one such model (Reuss and Johnson, 1985) will suffice. The model will demonstrate the insight into the processes involved that can be gained by their use. More complete descriptions of models of this type may also be found in Reuss (1980, 1983), Christopherson et al. (1982), and Chen et al. (1983). In its simplest form, the model consists of seven equations, including the CO_2–HCO_3^- equilibrium [Eq. (2-1)], the equilibrium expressions for the three reactions that control the concentrations of ionic aluminum species in solution [Eqs. (5-1), (5-2), and (5-3)], an equation describing the calcium-aluminum ion exchange equilibrium [Eq. (5-5)], the hydrolysis of water, and the charge balance equation.

$$Al(OH)_3 + 3H^+ \rightleftharpoons Al^{3+} + H_2O, \tag{5-1}$$

$$Al^{3+} + H_2O \rightleftharpoons Al(OH)^{2+} + H^+, \tag{5-2}$$

$$Al^{3+} + 2 H_2O \rightleftharpoons Al(OH)_2^+ + 2 H^+. \tag{5-3}$$

In the strict sense, Eq. (5-1) applies only to the dissolution of $Al(OH)_3$. The corresponding equilibrium expression can be written in logarithmic form as

$$\log K_{Al} = 3 \, pH - pAl, \tag{5-4}$$

where K_{Al} is the equilibrium constant for Eq. (5-1) and pAl is the negative logarithm of the molar activity of Al^{3+}. The concentration of these ionic aluminum species in solution at a given pH varies among soils, and in the model, this is accounted for by varying the value of K_{Al}. Although describing the processes of the dissolution of aluminum-bearing minerals in the soil by the simple dissolution of $Al(OH)_3$ is obviously a vast oversimplification, the constancy of $3 \, pH - pAl$ is predicted by H^+–Al^{3+} exchange considerations as well as by the solubility of gibbsite. Thus, Eq. (5-4) tends to hold for soils in which aluminosilicate minerals control aluminum solubility as well as those controlled by some form of gibbsite (Lindsay, 1979). If the controlling phase is an aluminosilicate mineral, the apparent value of K_{Al} is likely to be less than the 8.04 value commonly quoted for crystalline gibbsite (Lindsay 1979). Equation (5-4) is also useful in that it is a relatively simple matter to obtain an experimental estimate of log K_{Al} for any soil. The equilibrium constants for Eqs. (5-2) and (5-3) are relatively straightforward and are given by Lindsay (1979) as $10^{-5.02}$ and $10^{-9.30}$, respectively.

The expression of Gaines and Thomas (1953) has been chosen to describe the calcium–aluminum exchange system in this model (Reuss 1983).

$$K_s \cdot \frac{(Al^{3+})^2}{(Ca^{2+})^3} = \frac{E_{Al}^2}{E_{Ca}^3}. \tag{5-5}$$

The parentheses denote molar activities in the solution phase, K_s is the selectivity coefficient, and E_{Ca} and E_{Al} are the fractions of calcium and aluminum, respectively, on the exchange complex, expressed in equivalents per equivalent of exchange capacity charge. For a limited system in which calcium is the only base cation, E_{Al} can be taken as equal to $1 - E_{Ca}$. The distribution of cations in solution at any base saturation E_{Ca} is sensitive to the value chosen for the selectivity coefficient K_S. Although the available information concerning appropriate values of this coefficient is limited, the sensitivity of the model to changes in the value of K_S indicates that the characteristics that determine the value of this coefficient in any soil are important factors in determining the sensitivity of the soil and associated water to acid deposition impact.

The input requirements include the appropriate equilibrium constants, the CO_2 partial pressure of the soil atmosphere, the calcium–aluminum status of the soil exchange phase (i.e., the base saturation E_{Ca} and E_{Al}) and the concentration of SO_4^{2-} and Cl^- in solution. By solving the seven equations simultaneously, the model then calculates the concentrations of OH^-, HCO_3^-, H^+, Ca^{2+}, Al^{3+}, $Al(OH)^{2+}$, and $Al(OH)_2^+$ in solution.

5.3 Conceptualizing the Model

The model as presented in Sect. 5.2 consists of a set of equations. When values of certain parameters are input, the equations can be solved simultaneously to obtain the values of the remaining parameters. Although these solutions are extremely useful, the output can be best understood if the reader has a conceptual understanding of certain key relationships in the model. One of these, the CO_2–HCO_3^- equilibrium, has been discussed in Chap. 3. In this section, we discuss the interrelationships among Ca^{2+}, Al^{3+}, and H^+ that are implied by Eqs. (5-4) and (5-5). The constancy of 3 pH $-$ pAl [Eq. (5-4)] simply states that the activity of Al^{3+} in solution will be proportional to the cube of the activity of H^+ [Eq. (5-6) and Fig. 5.1].

$$(Al^{3+}) = K_{Al} (H^+)^3. \tag{5-6}$$

Values of log K_{Al} encountered in soils may range from <8.0, (i.e., undersaturated with respect to gibbsite) to >9.66, which is the expected value if the

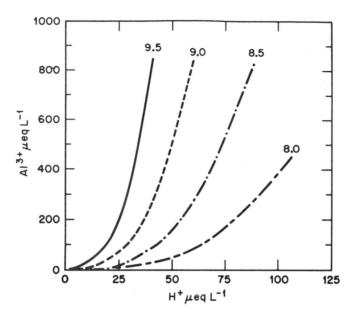

Figure 5.1. The relationship between Al^{3+} and H^+ for 3 pH − pAl values in the range of 8.0 to 9.5. Calculations assume activity coefficients of 0.96 and 0.70 for H^+ and Al^{3+}, respectively.

controlling mineral phase is amorphous $Al(OH)_3$ (Lindsay 1979). If the value of log K_{Al} is relatively high (e.g., 9.5), we would not expect the pH to drop below about 4.5 (32 μeq H^+/L), because further inputs of H^+ would only tend to bring more Al^{3+} into solution. At lower values of K_{Al}, this buffering effect would occur at lower pH and be less abrupt. Where gibbsite (log K_{Al} = 8.04) is the controlling phase, the pH in solution could drop to 4.0 or less. This is the aluminum buffering mechanism referred to by Ulrich (1983) as occurring below pH 4.2.

In soils, the total amount of exchangeable cations is usually much greater than the amount in solution at any one time. Therefore, exchange reactions that occur in response to changes in solution concentration generally will not significantly affect the exchange pool size, unless they occur repeatedly over a long period of time. Significant changes in exchangeable cations resulting from acid deposition would require years or decades. Therefore, for describing the short-term effects of changing solution concentration, we can regard E_{Al} and E_{Ca} as constant, and rearrange Eq. (5-5) to obtain

$$(Al^{3+}) = K_T (Ca)^{3/2} \qquad (5-7)$$

where

$$K_T = \left\{ \frac{E_{Al}^2}{K_s E_{Ca}^3} \right\}^{1/2} . \qquad (5-8)$$

Equation 5-7 states that the activity of Al^{3+} will be proportional to the 3/2 power of the Ca^{2+} activity. The relationship for two values of log K_S is shown in Fig. 5.2. The Al^{3+}–Ca^{2+} balance in solution is determined by the fraction of exchange sites occupied by Ca^{2+} and by Al^{3+} and the value of the coefficient K_S. For the values of the exchange coefficients shown, solutions will be dominated by Ca^{2+} if more than about 20% of the exchange sites are occupied by Ca^{2+}. Further reduction in calcium saturation will result in a rapid increase in solution Al^{3+}.

The percentage of calcium saturation referred to here is the calcium saturation at the pH of the soil. Because many soils contain significant amounts of pH-dependent charge, in acid soils, the cation exchange capacity (CEC) at soil pH is often substantially lower than the CEC that will be measured using an exchange solution buffered at pH 7.0. Thus, calcium saturation (i.e., base saturation) values based on CEC values determined at pH 7.0 are likely to underestimate the calcium saturation at soil pH.

Another important point is that at low calcium saturation, an increase in total solution concentration will cause the Al^{3+}/Ca^{2+} ratio in solution to increase. Therefore, an increase in solution concentration resulting from acid deposition will not only tend to increase the total cations in solution but will also increase the Al^{3+}/Ca^{2+} ratio. Whether or not this shift will cause a significant increase in Al^{3+} depends on the Al^{3+}–Ca^{2+} balance on the exchange sites and on the value of K_S for the particular soil.

From a mathematical standpoint, specifying the Al^{3+}–H^+ and Al^{3+}–Ca^{2+} relationships [Eqs. (5-4) and (5-5), respectively] also fixes the Ca^{2+}–H^+ relationships so that there is no separate equation for that purpose in the model. However, from a conceptual standpoint, it is useful to examine this relationship as well. Combining Eqs. (5-6) and (5-7), we obtain

$$(Ca^{2+}) = (H^+)^2 \left\{ \frac{K_{Al}}{K_T} \right\}^{2/3}. \tag{5-9}$$

By taking the negative log of Eq. 5-9 we obtain,

$$pH - 1/2 \ pCa = 1/3 \ \log \left\{ \frac{K_{Al}}{K_T} \right\}. \tag{5-10}$$

or

$$pH - 1/2 \ pCa = K_L, \tag{5-11}$$

where

$$K_L = 1/3 \ \log \left\{ \frac{K_{Al}}{K_T} \right\}. \tag{5-12}$$

The constant K_L is known as the lime potential (Schofield and Taylor 1955). It has long been used by soil scientists as an indication of the capability of the soil to maintain solutions dominated by Ca^{2+} and Mg^{2+} rather than H^+ or Al^{3+}. Although it is sometimes considered to be a Ca^{2+}–H^+ exchange coefficient, the

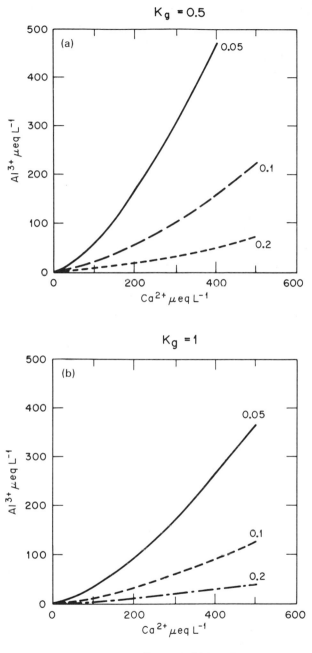

Figure 5.2. The relationship between Al^{3+} and Ca^{2+} in soil solution with the fraction of exchange sites occupied by Ca^{2+} ranging from 0.05 to 0.20. The log of the Gaines-Thomas exchange coefficient is 0.5 in (a) above and 1.0 in (b) below. Activity coefficients in solution are assumed to be 0.85 and 0.70 for Ca^{2+} and Al^{3+}, respectively.

derivation shown here does not assume a H^+–Ca^{2+} exchange process per se. From Eqs. (5-8) and (5-12), we see that K_L is a function of the solubility of aluminum in the system (K_{Al}), the Al^{3+}–Ca^{2+} selection coefficient (K_S), and the fraction of exchange sites occupied by Al^{3+} and by Ca^{2+} + Mg^{2+} (E_{Al} and E_{Ca}, respectively). Increasing the fraction of Ca^{2+} plus Mg^{2+} on the exchange complex will cause the lime potential to increase, as will an increase in either the value of either K_{Al} or K_S. Increases in the total solution concentration such as those likely to occur in response to acid deposition will not change the lime potential. However, such increases will increase the concentration of both H^+ and Ca^{2+}. The increase in Ca^{2+} will be proportionately greater than that of H^+.

The ion-exchange buffering effect referred to by Ulrich (1983) is apparent from Fig. 5.3. At a lime potential of 3.0 and 10 μeq H^+/L (pH 5.0), the solution will contain about 215 μeq Ca^{2+}/L. However, at 20 μeq/L (pH 4.70) the Ca^{2+} concentration must increase to about 860 μeq/L. At a lime potential of 2.5 the Ca^{2+} concentration at pH 5.0 and 4.7 would be 22 and 86 μeq/L, respectively. Note that an increase in total solution concentration will also cause an increase in the Ca^{2+}/H^+ ratio (and a decrease in the Ca^{2+}/Al^{3+} ratio). As in the case of Al^{3+}–Ca^{2+} exchange, an increase in total solution concentration will increase the proportion of the ion with the higher valence.

The buffering effect associated with high values of the lime potential is an important factor in determining the effect of acid inputs on soils and soil solutions. Soils having high or even moderate lime potential (perhaps 3.0 or more)

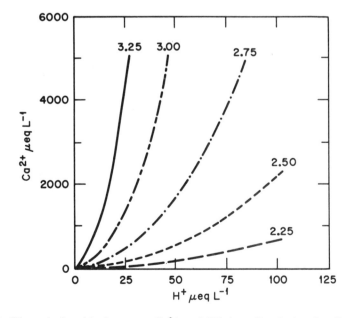

Figure 5.3. The relationship between Ca^{2+} and H^+ in soil solution for lime potential values in the range of 2.25 to 3.25. Calculations assume activity coefficients of 0.96 and 0.85 for H^+ and Ca^{2+}, respectively.

will be well-buffered against pH changes because of acid inputs. The effect of acid inputs in such soils would be largely to bring $Ca^{2+} + Mg^{2+}$ into solution. The sensitivity implications will be discussed more fully in Chap. 7.

5.4 Solution Concentration Effects

For the most part, we have chosen to present the output of this model as graphs of the calculated values of various parameters such as ion concentrations, pH, etc., as a function of the percentage of exchange sites occupied by the Ca^{2+} ion. In systems dominated by the divalent Ca^{2+} ion (actually $Ca^{2+} + Mg^{2+}$), this approximates the base saturation percentage. This method of presentation is useful in that, for a particular value of the input constants, it illustrates the probable status of soils having different base saturation levels. The lines should predict the nature of changes brought about by relatively small or moderate changes in base saturation. However, we do not intend to imply that if drastic changes in base saturation were brought about in any particular soil, the ensuing changes would follow these lines closely. Major changes in base saturation over a long period of time might be accompanied by depletion of the more-soluble aluminum-bearing minerals, so that the value of K_{Al} would change. Probably more important, the CEC is pH-dependent, decreasing as the H^+ concentration increases. This may be thought of as protonation of weak acid negative charges. Therefore, as the soil acidifies, the amount of bases required to maintain a given base saturation decreases. Finally, as the relative amounts of Ca^{2+} and Al^{3+} on the exchange complex change, a shift in the value of K_S may be observed (Pleysier et al. 1979). Whether this shift is the result of theoretical inadequacies of the Gaines-Thomas relationship [Eq. (5-5)] or changes in the character of the exchanger need not concern us here. A comparison of the Gaines-Thomas equation with some other commonly used ion-exchange relationships for the $Ca^{2+}-Al^{3+}$ exchange is given by Reuss (1983).

Figure 5.4a to d shows the distribution of ions in solution as a function of the percentage calcium saturation of the ion-exchange complex as calculated by this model. The "normal" condition is simulated by assuming a solution SO_4^{2-} concentration of 25 μeq/L, whereas the effect of H_2SO_4 and/or dry deposition is simulated by assuming a concentration of 250 μeq of SO_4^{2-}/L. A background of 20 μeq/L of $[Cl^- + NO_3^-]$ is assumed in both cases, as are values of 8.5 and 2.0 for log K_{Al} and log K_S, respectively, and a CO_2 partial pressure of 1%. Although these results are perhaps currently best regarded as semiquantitative, they nonetheless clearly illustrate the type of effects that can be expected from increased solution SO_4^{2-} levels.

Increasing the SO_4^{2-} concentration from 25 to 250 μeq/L decreases the concentration of HCO_3^- and increases the concentrations of cations in solution (i.e., Ca^{2+}, H^+, and the ionic aluminum species). The increased cation concentrations are required to maintain charge balance with the increased SO_4^{2-} in solution. When the input responsible for the increase in SO_4^{2-} is H_2SO_4, the soil reaction sequence would be either the reaction of H^+ with soil minerals to

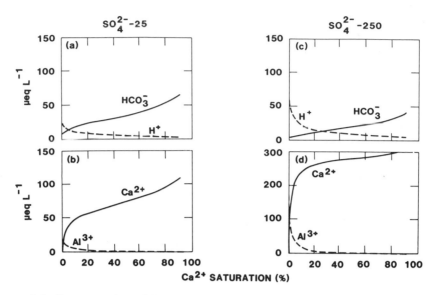

Figure 5.4. Concentration of individual ions as a function of percentage of calcium saturation as predicted by the chemical equilibrium model. Aluminum represents the sum of the ionic aluminum species in μeq/L. See text for input assumptions. (a) H^+ and HCO_3^- at 25 μeq SO_4^{2-}/L. (b) Ca^{2+} and Al at 25 μeq SO_4^{2-}/L. (c) H^+ and HCO_3^- at 250 μeq SO_4^{2-}/L. (d) Ca^{2+} and Al at 250 μeq SO_4^{2-}/L.

form aluminum species [Eqs. (5-1), (5-2), and (5-3)], followed by displacement of Ca^{2+} on the exchange complex by aluminum species, or the direct exchange of H^+ for Ca^+ followed by mineral dissolution and the replacement of exchangeable H^+ by Al species. Because the reservoir of ions on the exchange complex is much larger than that in solution, the result would be very similar, at least in the short term, if the input causing increased SO_4^{2-} concentration were a neutral salt such as $CaSO_4$. In that case, the reactions would be reversed (i.e., Ca^{2+} would displace aluminum from the exchange sites). If the resultant increase in aluminum species in solution is sufficient to exceed the solubility of aluminum in the system, some precipitation would occur, releasing H^+ ions [the reverse of Eq. (5.1)] and increasing aluminum solubility.

The H^+ lines from Fig. 5.4a and c have been converted to pH and plotted on the same graph for easy comparison (Fig. 5.5). In this example, the difference between the normal case (25 μeq SO_4^{2-}) and the acid deposition case (250 μeq SO_4^{2-}/L) is on the order of 0.3 pH units, although this difference is not constant at all base saturation levels. The relationships are affected by the values used for K_{Al} and K_S, so these results must be regarded only as examples. The effect of increasing solution concentration as a result of SO_4^{2-}, however, is always to depress pH as long as the soil has a net negative charge. This pH depression is known as the "salt effect" and is not the result of the acidity of the input solution.

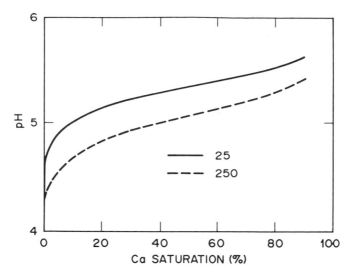

Figure 5.5. Soil solution pH as a function of percentage of calcium saturation at 25 and 250 μeq SO_4^{2-}/L as calculated by the chemical equilibrium model.

In order to satisfy the CO_2–HCO_3^- equilibrium as shown in Eq. (2-1), the increase in H^+ resulting from increased SO_4^{2-} must be accompanied by a decrease in HCO_3^-. This has the effect of reducing the bicarbonate alkalinity of the water that percolates through the soil, which in turn can have important consequences on the quality of surface waters, a situation that will be considered in more detail in Chap. 7. The HCO_3^- concentrations are also highly sensitive to the CO_2 partial pressure in the soil. The effect of increased CO_2 is to increase HCO_3^-, particularly at high base saturation. The slopes of the HCO_3^- lines in Fig. 5.4a and c are determined largely by the CO_2 partial pressure in the soil and can be expected to vary markedly from site to site and with time at a particular site. However, a depression associated with the increased solution concentration will be observed at all CO_2 levels.

Except at very low base saturation levels, the dominant cation is Ca^{2+} ($Ca^{2+} + Mg^{2+}$), particularly in the normal case in which the Ca^{2+} line tends to track the HCO_3^-, albeit at a slightly higher level reflecting the small background of strong acid anions (Fig. 5.4a and b). The transition between Ca^{2+}-dominated solutions and those dominated by H^+ and aluminum species occurs at low calcium saturation levels and is quite abrupt. High concentrations of ionic aluminum in solution are rare in natural systems because strong acid anions are generally scarce, and HCO_3^- concentrations are low at the H^+ concentrations required to bring ionic aluminum into solution.

The increase in ionic aluminum concentration resulting from increased SO_4^{2-} associated with acid deposition can be substantial, as can be seen by comparing values for aluminum in Fig. 5.4b and d. In this example, at 5% calcium saturation and 25 μeq SO_4^{2-}/L, the total charge associated with ionic

aluminum species is 5.4 μeq/L, with some 42% of this charge borne by monomeric Al^{3+}. Increasing the SO_4^{2-} to 250 μeq SO_4^{2-}/L increases the total charge
associated with aluminum species to 39.3 μeq/L, with >70% of the charge
associated with Al^{3+}. In the simulation, this increase in solution aluminum was
observed even though the fraction of exchange sites occupied by both calcium
plus magnesium and aluminum is the same in both cases. The increase in
aluminum is, therefore, the result of soil-solution reactions brought about by
the increased solution concentration and is not the result of any reduction in
exchangeable bases.

The relationship between the lime potential of the soil solution and the
fraction of the exchange sites occupied by calcium plus magnesium as calculated by the model is shown in Fig. 5.6. It is perhaps best understood from Eq.
(5-13) (Turner and Clark 1964; Reuss 1983). This equation can be obtained by
substituting into Eq. (5-10) the definition of K_T from Eq. (5-8) and rearranging

$$pH - 1/2\ pCa = 1/3\ \log K_{Al} + \log K_S + 1/6 \log \left(\frac{E_{Ca}^{\ 3}}{E_{Al}^{\ 2}}\right). \qquad (5\text{-}13)$$

When expressed in this way, it is apparent that the shape of the curve is fixed
by the term $1/6 \log (E_{Ca}^{\ 3}/E_{Al}^{\ 2})$, where E_{Ca} is the fraction of the exchange sites
occupied by Ca, and E_{Al} the fraction occupied by aluminum. Changing values
of $\log K_{Al}$ or $\log K_S$ (here fixed at 8.5 and 2.0, respectively) serve only to cause
a vertical displacement of the curve but do not change the shape. Therefore, at
least in soils in which the dominant exchangeable ions are Ca^{2+} and Mg^{2+}, a
single determination of the base saturation and the lime potential will define the
entire curve.

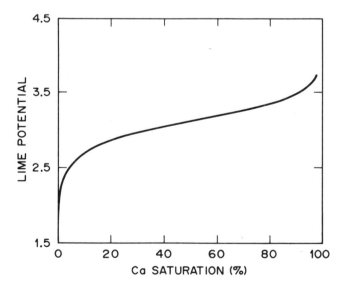

Figure 5.6. Lime potential (pH $-$ 1/2 pCa) calculated from Eq. (5.7), assuming $\log K_{Al}$
and $\log K_s$ equal to 8.5 and 2.0, respectively.

The shape of these curves corresponds well with the empirical curves reported by Clark and Hill (1964) for a range of soils from northeastern North America. The relatively flat portion of the curve corresponds to the ion-exchange buffer range discussed by Ulrich (1983).

In some models of this type (Reuss 1978, 1980; Christopherson et al. 1982;) the ion-exchange effect is included by assuming a constant lime potential. The advantage of the use of an exchange equation such as Eq. (5-5) rather than the assumption of a constant lime potential is that once K_{Al} and K_S are known, the model can be used for any base saturation, or if desired, the model can be modified to calculate changes in base saturation resulting from deposition. Although using the lime potential system has the advantage that K_L is easily determined experimentally, if K_{Al} and K_L are known, K_S can be estimated by using Eq. (5-13). A model based on the assumption of a constant lime potential such as that of Reuss (1980) or Christopherson et al. (1982) could easily be modified to accommodate changes in base saturation occurring as a result of cation depletion by adding the relationship shown in Eq. (5-13).

The nature of this transition from calcium- to aluminum-dominated solutions has several important implications concerning the effect of acid deposition. In the first place, as shown in the previous paragraph, substantial increases in ionic aluminum species are possible without prior reduction in exchangeable base cations as a result of cation export. This increase can occur as soon as the soil solution SO_4^{2-} concentrations respond to the impact, and conversely, if impact levels are reduced, the ionic aluminum will decrease along with the solution SO_4^{2-} levels. Secondly, the increase in solution aluminum that will occur as a result of increased levels of solution SO_4^{2-} at any calcium saturation level is sensitive to the values of K_{Al} and K_S for the particular soil. These parameters have generally not been considered when schemes for evaluating the sensitivity of soils to acid impact have been devised (McFee 1980; Wiklander 1980; Johnson 1981). Finally, the abruptness of the change in solution aluminum with calcium saturation has implications in determining the sensitivity of the soil to aluminum mobilization resulting from cation loss resulting from acid deposition. Cation export will have little effect on the calcium-aluminum balance in solution until calcium saturation is reduced to a low level. Further reductions in exchangeable calcium resulting from cation export will then result in a sharp rise in the concentration of ionic aluminum species, particularly Al^{3+}. Again, the actual base saturation at which this transition can be expected to take place depends on the values of K_{Al} and K_S. Whereas the solubility of aluminum in soils can vary from that of gibbsite by at least two orders of magnitude (i.e., values of log K_{Al} could vary from perhaps 6 to 10), from the data we have examined, most will probably fall within the range of 7 to 9.8. Although very little information is available concerning appropriate values of log K_S, Coulter and Talibuddeen (1968) reported values of 1.8, 2.1, and 3.1 for bentonite, illite, and vermiculite, respectively, and Bache (1974) reported values ranging from -0.9 to 3.1 for 11 soils from Great Britain. Limited model studies using a range of these constants suggest that the base saturation at which aluminum becomes a significant component of the solution could vary from <1 to 20% or even higher.

5.5 Cation Removal, $\Delta M/\Delta H$

Various authors (e.g., Wiklander and Andersson 1972; Krug and Frink 1983) have pointed out that in acid soils, base cation removal by acid deposition will be less than the H^+ input (i.e., $\Delta M/\Delta H < 1.0$, where ΔM is the cation removal and ΔH the H^+ input). $\Delta M/\Delta H$ is sometimes referred to as the "replacement efficiency" of H^+ for base cations. Krug and Frink (1983) have gone so far as to state that for acid rain inputs to soils below pH 5.0, $\Delta M/\Delta H$ will be much less than 1.0. For the system described by Fig. 5.4, ΔH can be taken as 225 $\mu eq/L$ (i.e., the difference in SO_4^{2-} in solution attributed to acid impact). ΔM is the difference in Ca^{2+} in solution (in μeq of Ca^{2+}/L) resulting from this increased SO_4^{2-}, (i.e., Ca^{2+} as shown in Fig. 5.1d minus that in Fig. 5.1b).

Three factors will tend to depress $\Delta M/\Delta H$ to values substantially below 1.0: (1) reduction of HCO_3^- in solution, (2) the increase of H^+ in solution, and (3) the increase in ionic aluminum species. The depression of HCO_3^- occurs mainly at high Ca^{2+} saturations, whereas the increase in H^+ occurs mainly at low Ca^{2+} saturation. In the previous example (Sect. 5.4), the reduction in the ratio resulting from the sum of these two effects ranges from 0.08 to 0.13 and is not much effected by the level of Ca^{2+} saturation. At low base saturation, the ionic aluminum increases to high levels (Fig. 5.4d), resulting in a corresponding decrease in $\Delta M/\Delta H$ as shown in Fig. 5.7.

In our example, values of $\Delta M/\Delta H$ substantially less than 1.0 occur only at very low base saturation (<15%), and the fact that soil solutions are often dominated by base cations even in areas where soils are acid, suggests that this

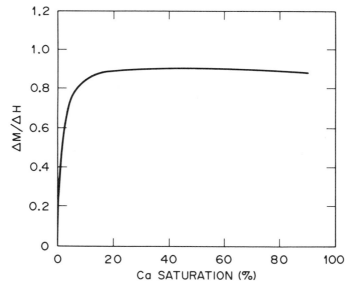

Figure 5.7. $\Delta M/\Delta H$ as a function of percentage of calcium saturation as calculated by the chemical equilibrium model.

will generally be the case. However, if soils having a low value of the aluminum–calcium exchange coefficient K_S are encountered, $\Delta M/\Delta H$ values may be considerably lower. Wiklander and Andersson (1972), found replacement efficiencies <0.5 with calcium saturation of 50% or more in a bentonite clay. Naturally acid soils such as the Spodosols commonly found in areas of heavy atmospheric deposition such as southern Scandinavia, eastern Canada, and the northeastern United States often have very low base saturation. Low values of $\Delta M/\Delta H$ could easily occur in such soils, either as an initial condition or a condition brought about by a relatively small amount of SO_4^{2-}-induced leaching resulting from deposition.

The occurrence of low values of $\Delta M/\Delta H$ must not be looked on as tending to preclude deleterious effects of acid deposition. Although it is true that low values of $\Delta M/\Delta H$ will be associated with rates of base cation loss substantially less than the H^+ loading, this reduced base cation loss is attained only at the expense of increased H^+ and ionic aluminum species in solution. This increase in aluminum, particularly Al^{3+}, is particularly worrisome in view of possible toxicities to vegetation within the ecosystem, as well as reduced alkalinity and toxicities to aquatic biota when discharged in the drainage waters.

5.6 Complexes and Precipitates

The possibility that the concentrations of the major ion species in the soil solution might be altered significantly by the formation of complexes or by the precipitation of other solid phases must be considered. For example, aluminum is strongly complexed by F^-, forming a number of species including AlF^{2+}, AlF_2^+, AlF_3°, and AlF_4^-. At the aluminum concentrations expected in acid soils, we can assume that any F^- in the system would be entirely complexed. For example, the reaction for the formation of AlF^{2+} can be written

$$Al^{3+} + F^- \rightleftharpoons AlF^{2+}. \tag{5-14}$$

The equilibrium constant for this reaction is given by Lindsay (1979) as $10^{6.98}$, so that the ratio of AlF^{2+} to F^- is given by

$$\frac{(AlF^{2+})}{(F^-)} = (Al^{3+}) \cdot 10^{6.98}. \tag{5-15}$$

At an Al^{3+} activity of $10^{-5.0}$ M (about 38 μeq Al^{3+}/L, assuming an activity coefficient of 0.79 as calculated by our model), the ratio $(AlF^{2+})/(F^-)$ is about 95. Therefore, in acid soils any F^- present would be almost completely complexed. Because the concentration of Al^{3+} is fixed by the pH and the solubility of the aluminum minerals, the result would most probably be an increase in the total aluminum in solution. If the F^- concentration were buffered by the dissolution of an F^--bearing mineral such as flourite (CaF_2), the formation of these complexes could prevent elevated Al^{3+} concentrations. Presumably, either minerals such as CaF_2 would not persist in acid soil or the rate at which they were exposed by weathering to reaction with the soil solution would be low, thus limiting F^- concentrations in solutions. Although aluminum complexed by

F^- would not be in the toxic Al^{3+} form, it would contribute to the titratable acidity of the drainage water. Therefore, these complexes would tend to buffer against pH rise when the soil solutions are released to the surface water (Chap. 7) because a rise in pH would decrease the concentration of Al^{3+} in solution, thus decreasing the ratio of AlF^{2+} complex to Al^{3+} in solution [Eq. (5-15)].

The Al^{3+} ion also forms complexes with SO_4^{2-}. Given the solution concentrations expected in acid rain affected soils, the species most likely to form in significant amounts is $Al(SO_4)^+$:

$$Al^{3+} + SO_4^{2-} \rightleftharpoons Al(SO_4)^+. \tag{5-16}$$

The equilibrium constant for this reaction is given by Lindsay (1979) as $10^{3.20}$, from which we can calculate that at the SO_4^{2-} concentration of $10^{-3.95}$ M, as used in our model for the acid-deposition-affected case, the $Al(SO_4)^+/Al^{3+}$ ratio would be about 0.18 and the $Al(SO_4)^+/SO_4^{2-}$ ratio would be 0.16. The significance associated with a fraction of the aluminum and SO_4^{2-} being bound together in a complex of this type is probably minimal. Comparison of results from model runs where $Al(SO_4)^+$ was not included with runs from a version modified to include this complex suggest that for our purposes the complex can be neglected. For example, at 1% calcium saturation inclusion of the complex reduced the predicted Al^{3+} in solution from 36.4 to 33.9 μmol/L, whereas the complex contributed 4.7 μmol/L aluminum to the solution. The total of all aluminum species in solution with and without considering the complex was 52.1 and 50.5 μmol/L, respectively. Inclusion of the complex reduced the predicted Ca^{2+} in solution from 48.6 to 46.4 μmol/L, apparently largely because of the reduction in total solution strength resulting from complexing of SO_4^{2-}. As calcium saturation increased so that the solutions were dominated by Ca^{2+}, the inclusion of the complex had virtually no effect.

Several authors (e.g., Adams and Rawajfih 1977; Adams and Hajek 1978; Nordstrom 1982; Prenzel 1983) have suggested the precipitation of minerals containing both aluminum and SO_4^{2-} as an alternative to the SO_4^{2-} adsorption mechanism to account for SO_4^{2-} retention in soils. The minerals most likely to be of interest in this regard include alunite $[KAl_3(OH)_6(SO_4)_2]$, basaluminite $[Al_4(OH)_{10}SO_4 \cdot 5H_2O]$, and jurbanite $[Al(OH)SO_4 \cdot 5H_2O]$ (Nordstrom 1982).

The precipitation of such minerals could theoretically serve as a control on the concentration of both Al^{3+} and SO_4^{2-} in the system. Up to this point, we have assumed that the control of Al^{3+} could be described by the constancy of 3 pH $-$ pAl, whereas the SO_4^{2-} retention or release could be described by an adsorption isotherm. It is therefore important to examine these reactions not only as a mechanism for the retention of SO_4^{2-} and H^+ in the system but also for implications concerning the validity of these assumptions.

The dissolution reactions and the associated log K° values for these minerals are:

Jurbanite (log K° = -17.2)

$$Al(OH)SO_4 \cdot 5H_2O \rightarrow Al^{3+} + SO_4^{2-} + OH^- + 5H_2O. \tag{5-17}$$

Basaluminite (log $K° = -117.5$)

$$Al_4(OH)_{10}SO_4 \cdot 5H_2O \rightarrow 4\ Al^{3+} + SO_4^{2-} + 10\ OH^- + 5\ H_2O. \quad (5\text{-}18)$$

Alunite (log $K' = -85.4$)

$$KAl_3(OH)_6SO_4 \rightarrow K^+ + 3\ Al^{3+} + 2\ SO_4^{2-} + 6\ OH^-. \quad (5\text{-}19)$$

The log $K°$ value for alunite is taken from Adams and Rawajfih (1977) whereas those for jurbanite and for basaluminite are from van Breemen (1973) and Singh and Brydon (1970), respectively.

The solubility diagram shown in Fig. 5.8 was constructed from the above log $K°$ values. As noted by Nordstrom (1982), at pSO$_4$ of 4.0 (about 230 μeq SO$_4^{2-}$/ L, assuming an activity coefficient of 0.85 for the divalent ion), alunite is the most stable of these minerals and below pH 4.5 is more stable than gibbsite. Although the solubility of alunite also depends to some extent on the activity of K$^+$ in solution, it is less sensitive to K$^+$ activity than to that of SO$_4^{2-}$. Basaluminite is less stable than alunite over the entire range and is also less stable than gibbsite. As shown in Fig. 5.8, jurbanite would become more stable than gibbsite at about pH 3.64. The solubility of jurbanite, however, is much more sensitive to the activity of SO$_4^{2-}$ in solution than is that of either basaluminite or alunite.

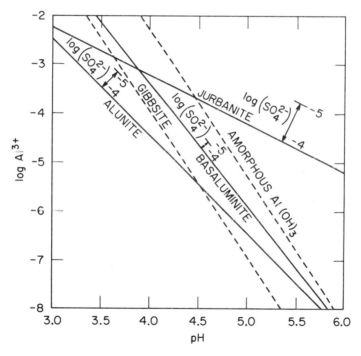

Figure 5.8. Solubility diagram to alunite, KAl$_3$(OH)$_6$(SO$_4$)$_2$; basaluminite, Al$_4$(OH)$_{10}$SO$_4$; jurbanite, Al(OH)SO$_4$; gibbsite, Al(OH)$_3$; and amorphous Al(OH)$_3$. The alunite line assumes the activity of K$^+$ to be 10^{-4} M.

The fact that the jurbanite and basaluminite lines lie above the gibbsite line in Fig. 5.8 does not preclude the formation of these phases, because the solubility of Al^{3+} exceeds that of gibbsite in many soils. However, the range of solution Al^{3+} will generally lie below the solubility of amorphous $Al(OH)_3$. Apparently basaluminite could precipitate in soils in which Al^{3+} levels are somewhat above that in equilibrium with gibbsite but would be thermodynamically less stable than alunite. The jurbanite line crosses that of amorphous Al^{3+} at about pH 4.5, suggesting that jurbanite could precipitate in some soils in which SO_4^{2-} levels are high. Jurbanite would also be thermodynamically less stable than either basaluminite or alunite.

A critical question relative to the possible formation of these compounds in soils affected by acid deposition concerns the degree to which their formation might serve as a control on the concentration of either Al^{3+} or SO_4^{2-} in solution. In the preceding model, we have assumed the constancy of 3 pH − pAl. If the formation of any of these compounds were controlling the Al^{3+} concentration, this assumption would no longer be valid. If jurbanite or basaluminite were controlling Al^{3+} concentration, the value of 3 pH − pAl would be proportional to pSO_4 + 2 pH. Assuming log $K°$ values as shown above, the relationship for jurbanite would be

$$3 \text{ pH} - \text{pAl} = -3.2 + pSO_4^{2-} + 2 \text{ pH}, \tag{5-20}$$

and for basaluminite,

$$3 \text{ pH} - \text{pAl} = 5.62 + 1/4 \,(pSO_4 + 2 \text{ pH}). \tag{5-21}$$

If alunite were the controlling phase, the Al^{3+} concentration would also depend on the K^+ concentration, so that

$$3 \text{ pH} - \text{pAl} = 0.47 + 1/3 \,(pK + 2 \, pSO_4 + 3 \text{ pH}). \tag{5-22}$$

From the limited information available, it seems unlikely that either alunite or basaluminite will control the concentration of Al^{3+} in soils affected by acid deposition. Although Adams and Rawajfih (1977) demonstrated that the $K°$ values for these compounds remained reasonably constant when various levels of K_2SO_4 were added to a Lucedale soil (initial pH 4.4), the observed constancy depended entirely on the variation in SO_4^{2-} activity, the Al^{3+} activity remaining nearly constant. In these experiments, 3pH − pAl also remained constant, indicating that if these minerals are precipitating, they are not serving as a control on the Al^{3+} concentration. Furthermore, when these minerals are formed by reacting $Al_2(SO_4)_3$ solutions with KOH or $Ca(OH)_2$, the value of 3 pH − pAl remained constant, apparently because of the formation of a gibbsite-type mineral.

Even though basaluminite or alunite were not controlling Al^{3+} concentration, their formation could be serving as a control on SO_4^{2-}. However, we consider this also to be unlikely. For example, in the results of Nilsson and Bergkvist (1983) shown in Fig. 5.9, many of the points lie in the region that would be supersaturated with respect to basaluminite. Alunite is not shown on this figure, and the axes are not appropriate for its inclusion. However, at

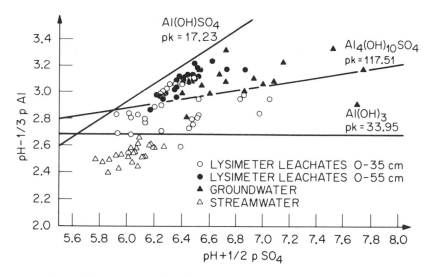

Figure 5.9. Equilibrium diagram for gibbsite, $Al(OH)_3$, jurbanite, $Al(OH)SO_4$, and basaluminite, $Al_4(OH)_{10}SO_4$. The data points are from the Lake Gordsjön site on the Swedish coast. (From Nilsson S. I. and Bergkvist B. *Water Air Soil Pollut.* 20:311–329, 1982. Reprinted by permission.)

similar SO_4^{2-} activity, it is less soluble than basaluminite, indicating that these solutions are probably also supersaturated with respect to alunite. Although it seems unlikely that these minerals are controlling SO_4^{2-} or Al^{3+} concentrations, they could be accumulating to some unknown extent in soils affected by acid deposition and, thus, forming an additional reservoir of SO_4^{2-}.

It is possible that jurbanite or a similar mineral may be serving as a control on the concentrations of SO_4^{2-} and/or Al^{3+}. For example, none of the points of Nilsson and Bergkvist (1983) as shown in Fig. 5.9 exceed the solubility assigned to jurbanite. The results of Prenzel (1983) also suggest that a mineral of this type is accumulating in the soils of the Solling site in Germany. An intriguing aspect of these results is that although the samples from under the spruce forests are consistent with a log $K°$ of -17.23 as we have used here, very few samples from the soils under beech forests would exceed a log $K°$ of -18.0.

The question of SO_4^{2-} adsorption versus precipitation of these or similar minerals cannot be completely answered at this time. Although some of the results shown here seem to support precipitation, many are probably equally consistent with an adsorption mechanism. For example, the results used by Adams and Rawajfih (1977) to support the constancies of the log $K°$ values for alunite and basaluminite can also be used to generate a typical adsorption isotherm. Nordstrom (1982) suggests that adsorbed SO_4^{2-} may be converted to one of these mineral forms on aging. He further suggests the jurbanite may provide an upper solubility limit, whereas alunite provides a lower limit. Both of these hypotheses would appear worthy of further investigation.

5.7 Organic Anions

Up to this point, our model has considered only the mineral anions. Organic anions, however, are known to play an important role in many soils. Because we do not now have a version of the equilibrium model that includes organic anions, it is necessary to rely on a subjective understanding of how these ions are likely to affect the system.

The organic anions are negatively charged fragments or complex molecules of organic matter. Organic acids are thought to be produced by partial degradation of lignin or synthesized by exoenzymes of certain soil microorganisms (Schnitzer and Khan 1972). These substances can contribute a substantial acidity to soil solutions and discharge waters. Waters from tundra or from boreal forest systems, lysimeter leachates from acid organic horizons, and drainage from acid peats are examples of acid solutions, often pH <4.0, in which the major anions are organic. Such solutions are invariably colored and, thus, are usually readily recognized in the field, although not all soil waters that are highly colored because of organic composition are acid. Besides the ability to support acid conditions in solutions, these materials readily form complexes with iron and aluminum. Solutions containing organic anions often contain much higher concentrations of aluminum than would be expected from the pH of the solution if only the ionic aluminum species were present. If the kinetics of dissolution are unfavorable, these complexes may result in solutions that are substantially undersaturated in Al^{3+} with respect to the controlling soil minerals, thus reducing biological toxicity. Also, these anions tend to become protonated as acidity increases, thus limiting the degree of acidity that can be supported.

Water moving downward through the organic layers of acid forest soils may become very acid even though mineral anions are very limited. When these acid (pH <4.0) solutions enter mineral horizons containing substantial amounts of iron and aluminum sesquioxides, the low pH tends to bring iron and aluminum into solution. The iron and aluminum are in turn chelated by the organic anions, lowering the activities of iron and aluminum in solution and causing yet more dissolution of iron and aluminum minerals. When the capacity of these anions to chelate or complex iron and aluminum is saturated, the complexes precipitate (Peterson 1976). The result is a horizon containing organic matter complexed with iron and aluminum; the drainage water from these horizons will be clear and the ability to support acidity resulting from organic anions is largely lost. The pH will generally be higher in and below this zone of precipitation than in the overlying organic horizons because of the loss of the organic anions and the consumption of protons in the process of mineral dissolution (Johnson et al. 1977; Ugolini et al. 1977). We can, therefore, expect that for soils and waters below this zone of precipitation of iron and aluminum, organic anions will have little effect on the processes we have described above. Above this zone, or in waters that escape from the soil without going through this precipitation process (i.e. colored waters), we must consider further the likely effects of these anions.

Krug and Frink (1983) have suggested that the incoming acid in acid deposition will protonate organic anions, so that the effect would simply be a replacement of organic ions by SO_4^{2-} ions, and, thus, there would be no net increase in anion strength. We find this unconvincing at best. In the first place, it could apply only to situations in which the solutions have not gone through the precipitation process described above. In the second place, it implies a direct replacement of organic anions by SO_4^{2-} and no accompanying increase in H^+ concentration. This, of course, would require that the titration curve of these acids be a "step function" (i.e., adding acid would only protonate the organic ions and not decrease the pH). This is inconsistent both with experience (Posner 1964) and with the diverse nature of these ions. Some pH change must occur to protonate these acids, and even a small change can considerably increase the concentration of ionic aluminum species in solution levels. Even more important, it seems that the process by which these ions lose negative charge is largely through complexing by iron and aluminum rather than direct protonation (Peterson 1976). On the other hand, it is entirely likely that, within horizons or drainage waters where organic anions are prevalent, the ability of these ions to complex aluminum and iron will mitigate the effects of acid deposition by reducing the concentration of ionic aluminum species in solution. At this point, we must consider that the degree to which this process will occur is largely speculation.

5.8 Summary

We can visualize the sequence of events that result from H_2SO_4 deposition as follows. The first response of the system to an increase in SO_4^{2-} deposition will be the initiation of a sulfate front moving downward through the soil. Above this front, the SO_4^{2-} concentration will rise to a value in approximate equilibrium with the new level of input. The rate of movement of the front will be controlled largely by the slope of the SO_4^{2-} isotherm for the soil, and this rate will not be much affected by the level of input, although biotic factors may modify the time required to reach a new equilibrium. The time required for the new level of input to be reflected in water that percolates through the soil and is later discharged to surface waters is highly variable, literally in the range of a few hours to many decades (Galloway et al. 1983).

With passage of this front, several effects can be expected. The concentration of the soil solution will increase because of the increase in SO_4^{2-}. There will be a concurrent slight drop in pH, but this change is unlikely to exceed 0.2 to 0.3 units and may be very difficult to observe experimentally. The concentration of HCO_3^- will also drop, but this will also be very difficult to measure experimentally because of the natural variability induced by varying CO_2 levels. The concentration of Ca^{2+} and/or ionic aluminum species will increase. The key to whether effects deleterious to the biotic components of the ecosystem will be found at this stage is in the relative increase in Ca^{2+} (and the other associated base cations), as compared to the increase in ionic aluminum species. This calcium–aluminum balance will, in turn, be a function of the base saturation E_{Ca} the solubility of the mineral phase controlling solution levels of aluminum K_{Al}, and the thermodynamic characteristics of the ion exchanger K_S. At high base saturations, perhaps >20%, the increase in aluminum in solution will usually be very small, probably unmeasurable, and not sufficient to adversely affect the biota. Whether significantly elevated aluminum will be observed at lower base saturation depends on the value of K_{Al} and K_S for the particular soil, parameters that have generally not been considered in schemes to define the sensitivity of the system.

Assuming that solutions are still dominated by bases, after passage of the SO_4^{2-} front the process of cation export begins. However, the mobilized base cations will remain available to plants as long as the sulfate front has not passed below the rooting zone, and in this stage, these cations are subject to the effects of vegetative recycling. Cation loss to the ecosystem begins when the SO_4^{2-} front passes below the root zone. Given common rates of acid deposition and existing levels of exchangeable soil bases in most soils, reduction of base saturation by this cation export process is likely to be slow, prompting many to discount the possibility of adverse effects from acid impact. In addition to the slowness of the process of export, release of cations as a result of the weathering of minerals will tend to offset the effects of cation loss. As a result, the possibility of deleterious effects is often considered to exist only in the case of soils having a combination of very low CEC and low base saturation.

To some extent we share this optimism. The acceleration of base cation removal resulting from acid deposition is a slow process, attenuated by mineral weathering effects. Significant removal of cations generally requires decades or even centuries. However, in soils having low base saturation, deleterious effects may be observed without cation export; aluminum mobilization may occur simply as a result of increase in solution concentration. We also call attention to the nature of the relationship between ionic aluminum species in solution and exchangeable base cations (i.e., the aluminum line in Fig. 5.4d). The transition between base-cation (i.e., Ca^{2+})-dominated systems and aluminum domination is abrupt, so that in a critical range, a relatively small change in base content resulting from acid-deposition-induced export could significantly increase the levels of aluminum in solution. Although such a shift would tend to reduce the rate of cation export because of a lowering of $\Delta M/\Delta H$, the consequences of aluminum mobilization in terms of deleterious effects on the biotic component and on water quality could be serious.

6. Forest Element Cycling

Up to this point, we have focused on the effects of acid deposition on soils and soil solution. Acid deposition may affect nutrient cycles in forests quite directly by changing fluxes via solution (i.e., foliage and soil leaching). These changes in solution flux may in turn result in changes in solid-phase fluxes (litterfall) and, thus, affect the rate and nature of nutrient cycles as well, as will be discussed below. Also, the mere presence of a forest canopy affects input via dry deposition (gaseous and particulate); coniferous forests generally receive higher dry deposition inputs because of their perennial canopy and greater leaf area (Höfken 1983; Matzner 1983; Skeffington 1983).

6.1 Definition of Terms

Schematic depictions of forest element cycling were presented in Fig. 1.1. The conceptualization of the cycle as well as the definitions of terms are to some degree based on the practicalities of measurement. In this model, we use definitions modified from Cole and Rapp (1981) to include root turnover:

uptake = nutrient return via litterfall + foliage leaching + root turnover,

requirement = nutrient content in new tissues (foliage + wood + roots), and

translocation = requirement − uptake.

6.2 Effects of Acid Deposition on Base Cation Flux

Clearly, acid deposition may directly affect foliar leaching, which in turn may affect either (1) litterfall or (2) total return or both. That is, (1) accelerated foliar leaching may reduce foliar cation concentrations such that litterfall return is reduced or (2) litterfall return may remain unaffected such that total return is increased. A consequence of the latter is that uptake is increased and translocation is reduced, the net effect being a shift from biochemical (or plant-internal translocation) to biogeochemical (litterfall-foliar leaching) cycling processes to satisfy requirement (the latter presumably remaining unchanged).

An additional reason that uptake might be increased is that solution concentrations of nutrients cation are increased by the presence of SO_4^{2-}. For these reasons, Johnson et al. (1985) speculate that the rates of K^+, Ca^{2+}, and Mg^{2+} cycling in eastern Tennessee deciduous forests have been accelerated by about twofold.

From the perspective of soil acidification, the only effect of nutrient cycling acceleration is in the distribution of acidification in the soil, not in the total magnitude. This can be seen in Figs. 6.1a through d. Figure 6.1a depicts an idealized, steady-state forest nutrient cycle to which we assign arbitrary, but reasonable, values [based on data from our experience (e.g., Johnson et al. 1985)] that serve as a conceptual baseline for considering the effects of H^+ deposition on base cation cycling and soil acidification. In the unaffected case, we have a small H^+ input (0.01 keq $\cdot$ ha^{-1} $\cdot$ year^{-1}) to the forest canopy and a net return of base cations via litterfall (2.0 keq $\cdot$ ha^{-1} $\cdot$ year^{-1}) and throughfall (0.2 keq $\cdot$ ha^{-1} $\cdot$ year^{-1}). The forest floor is in steady-state condition, and the soil is being acidified through natural leaching processes (i.e., leaching of cations in association with either HCO_3^- or organic anions). If acid deposition occurs and foliar leaching is thereby accelerated, two basic consequences are possible as discussed previously: (1) cation return and uptake rates increase or (2) return and uptake remain constant and the litterfall component of return decreases as foliage leaching increases. In the former case, illustrated in Fig. 6.1b, uptake compensates for accelerated cation leaching from the canopy by increasing from the baseline level of 2.2 keq $\cdot$ ha^{-1} $\cdot$ year^{-1} to a new level of 3.2 keq $\cdot$ ha^{-1} $\cdot$ year^{-1}. Thus, litterfall return to the forest floor is not affected, throughfall return is increased by 1.0 keq $\cdot$ ha^{-1} $\cdot$ year^{-1}, and total return is increased by 1.0 keq $\cdot$ ha^{-1} $\cdot$ year^{-1}. In short, the natural cycle is speeded up. In this case, the forest floor remains in steady-state condition because the accelerated leaching output is matched by accelerated cation return via throughfall. The mineral soil is acidified, however, since leaching (1.5 keq $\cdot$ ha^{-1} $\cdot$ year^{-1}) has increased by the same amount as the increase in H^+ input.

In the second case, depicted in Fig. 6.1c and d, the effect of accelerated canopy leaching is to reduce return via litterfall. This situation might conceivably lead to foliar cation deficiency levels, even in cases where soil supplies and plant uptake would otherwise be adequate to meet plant requirements. Rehfuess et al. (1982) postulate that this is the cause of deficiency levels of magnesium and calcium in foliage in forests in Germany that are currently undergoing

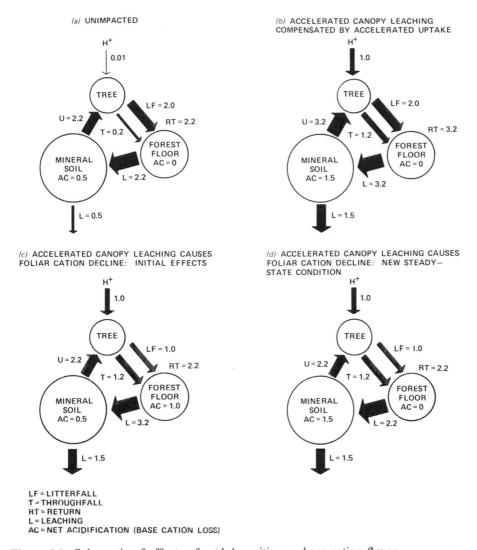

LF = LITTERFALL
T = THROUGHFALL
RT = RETURN
L = LEACHING
AC = NET ACIDIFICATION (BASE CATION LOSS)

Figure 6.1. Schematic of effects of acid deposition on base cation fluxes.

dieback and decline. Although total cation return is unaffected in this case, foliar exchange would initially be expected to have a net acidifying effect on the forest floor because the base cation content in litterfall is reduced. Cations reaching the forest floor via throughfall will undergo various exchange reactions, but with the presence of a mobile anion, total cation leaching from the forest floor will be greater than in the unaffected case (Fig. 6.1c). In this scenario, the net effect is that base output > input for the forest floor, and thus, the forest floor is acidified. At this stage, the mineral soil remains unaffected. Eventually, the forest floor will reach a new, lower steady-state condition, and the mineral soil will then be acidified as in the case shown in Fig. 6.1b. In both of the above cases (Fig. 6.1c and d), the net system loss of base cations is equal to the H^+ input (i.e., it is assumed that the associated anion is mobile). Because of the effects of mineral cycling processes, however, this acidification is initially manifested in different parts of the system (i.e., forest floor or mineral soil, depending on the changes caused in uptake and litterfall).

Perhaps the most important point is that, ultimately, the acidification of the foliage-litter-soil system from atmospheric acid deposition is not affected by canopy exchange processes (but may be enhanced by canopy dry deposition). Although the analysis suggests the possibility that an acidification of the forest floor may precede the acidification of the underlying mineral soil, the total acidification of the system depends on the excess of mobile anions over base cations in the flux from the atmosphere to the ecosystem. As in the fluxes shown in Fig. 6.1, the exchange of H^+ in precipitation for base cations in the foliage will increase the base cation content of the throughfall and decrease the H^+ ion concentration; that is, the pH of the throughfall will be increased relative to that of the precipitation. This pH increase has been observed by many workers, and is likely to be particularly evident in summer throughfall in deciduous forests (Lindberg et al. 1979; Richter et al. 1983). Although this increase in pH during passage through the canopy implies that the rainfall has been neutralized, it by no means implies that the capability to cause acidification of soils and drainage waters has been reduced or eliminated. This misconception is usually a result of an attempt to evaluate acidification potential in terms of H^+ ion fluxes, a procedure that is not necessarily incorrect, but it is easily misapplied because many of these fluxes are implicit in the associated chemical reactions and cannot be evaluated by simple pH measurements.

In the natural system, bases return from the foliage to the soil largely as salts of weak organic acids. In the throughfall, these may be comparatively simple salts such as malates, whereas in litterfall they may be negatively charged sites on larger organic polymers. In the event that these organic molecules are destroyed by biological respiration processes, the bases would exist as bicarbonates. As either salts of weak organic acids or bicarbonates, these bases contribute to the alkalinity (Chap. 7) of the system. If these bases are stripped away by a strong acid such as H_2SO_4 to form SO_4^{2-} salts, they no longer contribute to the alkalinity of the system, and the system has effectively been acidified, even though the pH of the throughfall is increased in the process. In the natural system, the capacity to buffer against acid inputs is stored as salts of organic acids or as HCO_3^- salts. To the degree that atmospheric H^+ input fluxes are accompanied by mobile strong acid anions, this buffering capacity is lost in the exchange of these base cations for the H^+ in atmospheric inputs.

6.3 Cation Nutrient Effects

In acid deposition effects studies, base cations are often lumped together and compared to H^+ and Al^{3+} in attempts to assess "base cation status" and acidity. Because trees take up cations individually and substitutions among individual cations are made only to a limited extent (if at all), lumping of base cations is an inadequate means of assessing cation nutrient effects.

Exchange relationships dictate that concentrations of divalent cations in soil solution will increase more in response to increased SO_4^{2-} concentrations than concentrations of monovalent cations. Thus, one could predict that Ca^{2+} and Mg^{2+} export from soils will be more readily affected by acid deposition than K^+ export from soils. However, it is not clear that the same exchange processes control foliar leaching. For instance, a much larger fraction of K^+ return typically occurs via crownwash than is the case for Ca^{2+} and Mg^{2+} (Cole and Rapp 1981). Also, there is a marked seasonality in foliar leaching in which peak values occur near senescence, just prior to litterfall (Henderson et al. 1977). Thus, it seems unlikely that exchange equations developed for soils will apply to foliar leaching. Nonetheless, several studies have shown that foliage leaching is accelerated to some degree by acid deposition to forest canopies (Wood and Bormann 1975; Abrahamsen 1980; Lee and Weber 1982). Soil cation leaching is unquestionably increased, and the rate of cation nutrient cycling may increase for the reasons discussed previously.

Not all base cations need show a net loss from soils by leaching even if total base cation leaching exceeds base cation input. For instance, Johnson et al. (1985) note net leaching losses of Mg^{2+}, K^+, and Na^+ but a net gain of Ca^{2+} in a chestnut oak forest having low calcium status in eastern Tennessee (Table 6.1). These forests accumulate considerable quantities of calcium in their woody biomass, and this may well constitute a conservation mechanism. That is, calcium accumulation in biomass may reduce soil exchangeable calcium, resulting in less calcium leaching relative to Mg^{2+}, K^+, and Na^+ (or, ultimately, Al^{3+}). The net gain of calcium can be the result of one or both of two factors: (1)

Table 6.1. Cation Budgets for a Chestnut Oak Forest in Tennessee and a Beech Forest in West Germany

	H^+	Al^{3+}	Ca^{2+}	Mg^{2+}	K^+	Na^+	Sum of
			($keq \cdot ha^{-1} \cdot year^{-1}$)				Bases
	Walker Branch, Tennessee—Chestnut Oak[a]						
Atmospheric input	1.36	b	0.41	0.11	0.02	0.06	0.60
Leaching	0.007	b	0.29	0.30	0.06	1.07	1.72
Balance	+1.35		+0.12	−0.19	−0.04	−1.01	−1.12
Vegetation increment		b	0.9	0.1	0.07		1.07
Balance		b	−0.78	−0.29	−0.11		−2.19
	Solling, FRG—Beech[c]						
Atmospheric input	1.39	0.20	1.33	0.35	0.24	0.63	2.55
Leaching	0.53	1.74	0.82	0.18	0.09	0.60	1.69
Balance	+0.86	−1.54	+0.51	+0.17	+0.15	+0.03	+0.86
Vegetation increment		0.07	0.39	0.45	0.26	0.03	1.13
Balance		−1.61	+0.12	−0.28	−0.09	0	−0.27

[a] From Johnson et al. (1985).
[b] Not routinely measured but assumed trivial in soil solution because of pH $\sim$ 6.0.
[c] From Ulrich (1980).

reduction of exchangeable Ca^{2+} by tree uptake to levels below steady-state (input = output) values or (2) increases in Ca^{2+} input. In the absence of forest uptake, leaching would presumably reduce exchangeable Ca^{2+} to a steady-state value (where input + weathering = output) but not below this. Calcium uptake by trees may reduce exchangeable Ca^{2+} to lower than steady-state values, resulting in a net system accumulation of Ca^{2+}, even while the soil is experiencing acidification by net leaching of K^+, Mg^{2+}, and Na^+ (Table 6.1). The accumulation of calcium as well as Mg^{2+} and K^+ by trees is also acidifying to the soil, however.

Data from the Solling site in West Germany, also shown in Table 6.1, illustrate a situation in which all base cations accumulate in the ecosystem (input > leaching output), and only with vegetation uptake does the soil base cation balance become negative (i.e., the soil is acidified), except in the case of Ca^{2+}, which shows net accumulation in the soil. Whether this ecosystem accumulation is the result of vegetation uptake, increased base cation inputs, or both is not known, but it illustrates that even heavily acid-deposited forest ecosystems need not experience net losses of base cations. This is not to say that base cation budgets have been or are now unaffected by acid deposition. If these soils were once in a state of higher base saturation, they could conceivably have been acidified (in part, at least) by acid deposition at these high rates (assuming weathering did not offset base cation loss). Also, exchange equations dictate that concentrations of all cations must be higher under these conditions to balance SO_4^{2-} than they would be if lower SO_4^{2-} concentrations prevailed. Finally, as we have mentioned previously, the conservation of base cations at this site is at the expense of a considerable amount of Al^{3+} leaching, with all of the attendant undesirable effects (Ulrich et al. 1980). The price of Al^{3+} mobilization is not paid (yet) by Ca^{2+} conservation at Walker Branch, however, because Mg^{2+}, K^+, and Na^+ are lost.

7. The Aquatic Interface

Soil-mediated effects are an important factor in determining the chemistry of surface waters even though (1) some waters do not intimately contact soil before entering lakes and streams and (2) soil-contacted waters often are further modified by contact with rocks and minerals after leaving soils and before being discharged to surface waters. The nature of soil-mediated effects is often not well understood. Some workers have been skeptical that acid deposition could cause significant acidification of surface waters that come into contact with soils because of the large buffering capacity inherent in soils through such processes as cation exchange, mineral dissolution, and protonation of organic acids (Krug and Frink 1983). Others have assumed that the combination of documented increases in acidity along with high SO_4^{2-} concentrations in areas of heavy acid deposition loading were sufficient a priori evidence to implicate acid deposition as the major causative agent (Schindler 1980; Schofield 1980). The resultant controversy has been enormous, as might be expected when apparently conflicting evidence is available on an issue that bears on regulatory policies that, in turn, could cost huge sums for compliance. It is unrealistic to expect that we can currently contribute sufficient insight into the relevant processes to resolve such a conflict. However, given our current level of understanding, perhaps this chapter can help to focus efforts on those processes that have a valid theoretical basis and, thus, help toward arriving at a rational resolution of the conflict.

In Chap. 5, we suggested a mechanism by which waters could become acidified by acid deposition loading without involving significant acidification of soils, at least in the sense of acidification resulting from cation export (also Reuss and Johnson, 1985). The development of this mechanism relied heavily on equilibrium models. In this chapter, we shall review and amplify these concepts and their implications for the aquatic system.

7.1 Alkalinity Concepts

An understanding of the processes involved as soil solution is released to become surface water relies heavily on the concept of alkalinity as commonly used in water chemistry. In its simplest form, alkalinity is defined as the sum of the bicarbonate (HCO_3^-), carbonate (CO_3^{2-}), and hydroxyl (OH^-) ion concentrations minus the hydrogen (H^+) ion concentration.

$$\text{alkalinity} = 2\,[CO_3^-] + [HCO_3^-] + [OH^-] - [H^+], \tag{7-1}$$

where the brackets indicate the molar concentrations. Various units may be used to express alkalinity, but for our purposes microequivalents per liter is most convenient. In the acidic systems with which we are concerned, CO_3^{2-} can usually be safely neglected. For simplicity we shall omit this ion from the remaining discussions.

Given Eq. (7-1), charge balance considerations also mandate a relationship between the remaining cations and anions. For illustration, consider water in which the cations are H^+, Ca^{2+}, Mg^{2+}, Na^+, and K^+, and the anions are OH^-, HCO_3^-, SO_4^{2-}, Cl^-, and NO_3^-. The charge balance equation for this system is given by

$$2\,[Ca^{2+}] + 2\,[Mg^{2+}] + [Na^+] + [K^+] + [H^+]$$
$$= [HCO_3^-] + [OH^-] + [Cl^-] + [NO_3^-] + 2\,[SO_4^{2-}], \tag{7-2}$$

which rearranges to,

$$2\,[Ca^{2+}] + 2\,[Mg^{2+}] + [Na^+] + [K^+] - [Cl^-] - [NO_3^-] - 2\,[SO_4^{2-}]$$
$$= [HCO_3^-] + [OH^-] - [H^+]. \tag{7-3}$$

The right side of Eq. (7-3) is the same as for Eq. (7-1) (neglecting CO_3^{2-}). Therefore, the left side of Eq. (7-3) also defines the alkalinity in terms of microequivalents per liter of positive charge rather than microequivalents per liter of negative charge, as in Eq. (7-1). Thus, alkalinity may also be defined as the sum of the concentration of base cations minus the concentration of strong acid anions; negative alkalinity values denote strong acid acidity.

One of the most useful aspects of the alkalinity concept is that, although both pH and HCO_3^- concentrations vary with the CO_2 partial pressure [Eq. (2-1)], in an isolated water system alkalinity is independent of CO_2 partial pressure. That this should be true is readily understood from Eq. (7-3), because the right-hand side contains only the concentrations of base cations and strong acid anions, neither of which are dependent on CO_2. The concentrations of the ions on the right side of Eq. (7-3) all vary with the CO_2 partial pressure but must vary in such a way as to maintain the equality of the sum.

The relationship among pH, CO_2 partial pressure, and alkalinity is important in determining the acidity of surface waters. This relationship is illustrated in Fig. 7.1, which is obtained by a simultaneous solution of Eqs. (7-1) and (2-1), and the equilibrium equation for the hydrolysis of water,

$$(H^+)(OH^-) = 10^{-14}. \tag{7-4}$$

The most striking characteristic of Fig. 7.1 is the tendency for the lines of equal alkalinity to converge as CO_2 partial pressures increase and to diverge as CO_2 decreases. Carbon dioxide partial pressures in soil are commonly in the range of 1 to 5% and higher on occasion, whereas CO_2 partial pressures in surface waters are likely to be near or perhaps slightly above the atmospheric level of 0.03% (Wright 1983). The pH of a water having an alkalinity of 10 μeq/ L at atmospheric CO_2 will be near 6.5, whereas the pH of water with alkalinity of -10 μeq/L will be about 5.0. Thus, a change in alkalinity of only 20 μeq/L can result in a pH change of 1.5 units at the CO_2 levels common to surface waters. In the range of CO_2 partial pressures commonly found in soils (i.e., 1 to 10%), the pH values of these same waters would both be near 5.0, and differences resulting from the alkalinity would be of the order of only 0.1 to 0.3 units.

From the above illustration, it is clear that large changes in soil solution pH are not necessary to account for very substantial pH changes in surface waters. In the critical range near zero alkalinity, small changes in soil solution pH can

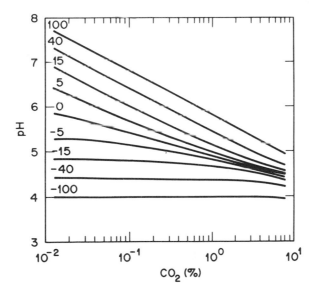

Figure 7.1. Effect of CO_2 partial pressure and alkalinity (μeq/L) on water pH.

result in very large changes in surface water pH when the soil solution is discharged to the lower CO_2 environment of the surface. To account for the ability of acid deposition to cause pH changes in surface waters, the scientific issue is not the absence of correspondingly large changes in soil solution pH, but rather, whether appropriate mechanisms exist whereby acid deposition could cause the requisite changes in alkalinity.

At least two important complexities arise when the alkalinity concepts are applied to surface waters that have been in contact with soils. The first of these is the presence in solution of the ionic aluminum species, Al^{3+}, $Al(OH)^{2+}$, and $Al(OH)_2^+$. In addition to the potential toxic effects of the monomeric Al^{3+} ion, the processes by which these ions are brought into and removed from solutions [Eqs. (5-1) to (5-4)] provides a pH buffering effect. Whether such buffering should be thought of as contributing to the acidity or the alkalinity of the system is an arbitrary decision that ultimately depends on the range of tolerance of species of interest to the biologist. A common approximation is to consider all aluminum species as acid so that alkalinity is defined by Eq. (7-5) (Henriksen 1980; Wright 1983; Reuss and Johnson, 1985).

$$\text{alkalinity} = [HCO_3^-] + [OH^-] - [H^+] - 3\,[Al^{3+}]$$
$$- 2\,[Al(OH)^{2+}] - [Al(OH)_2^+]. \tag{7-5}$$

This definition slightly underestimates the alkalinity, because some aluminum, particularly $Al(OH)_2^+$, remains in solution as the pH rises. In our example with log K_{Al}, log K_s, and CO_2 set at 8.5, 2.0, and 1%, respectively, the total acidity resulting from ionic aluminum species is calculated to be 3.3 μeq/L at pH 5.0 and 1.3 μeq/L at pH 5.5. From a practical standpoint, this error is probably less than the errors involved in measurement.

The second complexity arises when solutions have an appreciable component of nonbicarbonate alkalinity (i.e., weak acid anions other than HCO_3^- are present). An example of an inorganic anion of this type is the orthophosphate ion, PO_4^{3-}. However, orthophosphate concentrations are usually low enough to be neglected in natural waters. The most common source of nonbicarbonate alkalinity arises from organic acids, which are often present in solutions from forest litters or surface soils. In Spodosols, organic anions may be dominant in soil solutions to the depth of the eluvial horizon, often in excess of 30 to 40 cm. If alkalinity is determined in such solutions by titration with acid, the result includes both HCO_3^- alkalinity and alkalinity arising from the protonation of weak organic acids. If the alkalinity measured in this manner is interpreted as HCO_3^- alkalinity, the result will overestimate HCO_3^- alkalinity. It is possible to separate the organic and HCO_3^- components of alkalinity, as determined by titration, by estimating HCO_3^- independently. This can be done either by determining pH after equilibration with a known CO_2 partial pressure and calculating HCO_3^- using Eq. (2-1) or by determining pH and total inorganic carbon (i.e., HCO_3^- plus H_2CO_3), after which HCO_3^- may be calculated using the dissociation constant for H_2CO_3, which is given by Lindsay (1979) as $10^{-6.36}$.

7.2 Alkalinity in Soil Solution

The soil-mediated process by which surface waters may be acidified as a result of acid deposition is best understood by considering how, under pristine conditions, drainage waters from acid soils become nonacidic surface waters. As shown above, alkalinity of surface waters is independent of CO_2 partial pressure. One conceptual model of this independence is that, as CO_2 pressure causes the concentration of H_2CO_3 to increase, one H^+ and one HCO_3^- ion are formed for each H_2CO_3 molecule that dissociates, so that the total of ($HCO_3^- + OH^- - H^+$) remains constant. Although the alkalinity of surface waters is independent of CO_2, the alkalinity of soil solutions is highly dependent on CO_2 partial pressure. This difference arises from the influence of the cation-exchange system in soils. When increased dissociation of H_2CO_3 takes place in soil solutions as a result of increased CO_2, the sum of ($HCO_3^- + OH^- - H^+$) does not remain constant because most H^+ formed does not remain in solution. Instead, it tends to exchange with other cations on the exchange sites or dissolves soil minerals releasing aluminum, which then replaces Ca^{2+} (or other base cations) on the exchange complex. The result is that the new (H^+) (HCO_3^-) product required by the increased CO_2 partial pressure [Eq. (2-1)] must be satisfied largely by increased HCO_3^- rather than by an equal increase in H^+ and HCO_3^-, as would be the case in waters not in contact with the exchange system. This increased content of Ca^{2+} and HCO_3^- is reflected in increased alkalinity.

Thus, in the CO_2-enriched atmosphere of the soil, even in acid soils, the solutions may have positive alkalinities. When these solutions are released to the lower CO_2 environments of the surface waters, the pH increases as CO_2 degassing takes place (Fig. 7.1). The result is that, even in regions where soils are acid, in clear surface waters (i.e., those in which organic anions are low), the predominant anion is HCO_3^-, and the pH is typically >6.0 and may be 7.0 or higher.

The dependence of soil solution alkalinity on CO_2 partial pressure can be illustrated using the model described in Chap. 5. This effect is clearly evident in Fig. 7.2, in which the calculated alkalinity of soil solution is plotted against calcium saturation at 1, 3, and 5% CO_2. The assumptions for these plots are the same as those for the examples in Chap. 5 (i.e., log K_{Al} and log K_s equal to 8.5 and 2.0, respectively, and a background of 20 μeq/L ($Cl^- + NO_3^-$). The normal condition (1) is represented by a SO_4^{2-} concentration of 25 μeq SO_4^{2-}/L and the acid deposition affected case (2) by 250 μeq of SO_4^{2-}/L.

In Fig. 7.2, the 1% CO_2 case is actually the same as that used for the example shown in Fig. 5.4. Neglecting the very small amount of OH^- present in these systems, the 1% alkalinity line in Fig. 7.2a can be obtained by subtracting the sum of the H^+ plus aluminum shown in Fig. 5.4a and b, respectively, from the HCO_3^- in Fig. 5.4a. In addition to the effect of CO_2 on alkalinity that is apparent in Fig. 7.2 by comparing Fig. 7.2a and b, we note that at all CO_2 levels the effect of the increase from 25 to 250 μeq SO_4^{2-}/L is to depress alkalinity. This

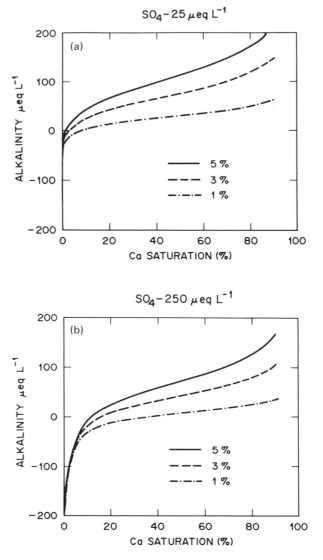

Figure 7.2 (a) Effect of percentage calcium saturation and CO_2 on alkalinity in soil solution as calculated by the chemical equilibrium model for 25 (b) and 250 μeq SO_4^{2-}/ L. See text for other input parameters.

depression is particularly noticeable at low base saturation. At 25 μeq SO_4^{2-}/L at 1, 3, and 5% CO_2, the alkalinities are positive above 8, 3, and 2% calcium saturation, respectively. At these same CO_2 levels and 250 μeq SO_4^{2-} in solution, the switch from negative to positive alkalinities occurs at 36, 17, and 14% calcium saturation, respectively. This depression of alkalinity by increased SO_4^{2-} occurs because of the decrease in pH caused by the salt effect (Figs. 5.5 and 7.3) and the concurrent decrease in HCO_3^- and increase in aluminum in

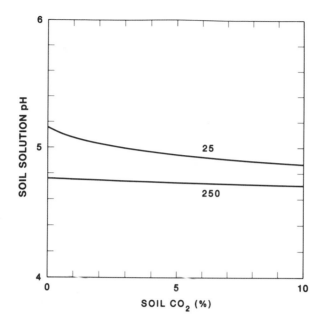

Figure 7.3. Effect of CO_2 on soil solution pH at 15% calcium saturation as calculated by the chemical equilibrium model for 25 and 250 μeq SO_4^{-2}/L. See text for other input parameters. (From Reuss J. O. and Johnson D. W. *J. Environ. Qual.* 12:263–270, 1985. Reprinted by permission.)

solution (Fig. 5.4). Note that in soil solutions, the effect of CO_2 on pH (Fig. 7.3) is small, particularly at the high SO_4^{2-} concentration, even though there is a marked effect of CO_2 on alkalinity (Fig. 7.2). In the field, CO_2 levels will be highly variable over time, so that alkalinity will also vary over time. The alkalinity of the waters will, thus, represent some integration of CO_2 levels that vary over both time and space.

To appreciate fully the significance of the depression of alkalinity by increased SO_4^{2-} on the acidity of surface waters, the effect on pH after degassing of CO_2 must be considered, rather than the effect on the pH of soil solution. The effect on soil solution pH is small, usually on the order of a few tenths of a pH unit. However, the effect on pH of the degassed water depends on the alkalinity (Fig. 7.1). If the alkalinity of the system unaffected by deposition is high (e.g., 100 μeq/L), the pH of the degassed water will be high and will not be much affected by a depression of alkalinity on the order of even 20 or 30 μeq/L. However, a depression of this magnitude, which resulted in a switch from positive to negative alkalinity, could result in a pH change in the surface water (of as much as 1.5 units) resulting from increased SO_4^{2-}.

The pH increase resulting from CO_2 degassing of acid solutions may be modified somewhat by the precipitation of aluminum compounds as the pH rises. The soil model we have described can be used to predict the effect of degassing of waters simply by deleting the ion-exchange equations from the model (Reuss and Johnson, 1985). The solubility of the aluminum compounds precipitated may not be the same as that of the compounds that are dissolving in the soil, so that appropriate K_{Al} values may be different for the water and soil systems. However, the pattern of equal alkalinity lines that is obtained from calculations of this type is not greatly different from that in Fig. 7.1 and, therefore, does not alter significantly the conceptual model obtained from the simpler system.

7.3 Naturally Acid Waters

Natural waters having substantial acidity do exist and are quite common in many areas. The discharge of acid drainage waters from soils requires a mobile anion, and although high concentrations of Cl^-, SO_4^{2-}, or NO_3^- occasionally may occur as a result of geological, geographic (coastal), or biological circumstance, in the overwhelming majority of cases, organic anions predominate in naturally acid waters. These waters occur in situations in which drainage from organic soil horizons is discharged directly into surface waters without passing through mineral horizons in which iron and aluminum minerals are dissolved and the organics precipitated out as iron and aluminum complexes. In general, waters dominated by organic acids are much less toxic to biological communities than those dominated by mineral acids, apparently because such waters retain a capacity to complex iron and aluminum in solution and, thus, depress the concentration of ionic species, particularly Al^{3+}.

A good deal of controversy has occurred over the question of natural versus anthropogenic water acidification, much of it apparently stemming from the use of pH to define acidity without a more complete characterization to separate organic-dominated from SO_4^{2-}-dominated waters. Unfortunately, this situation is complicated by analytical problems that occur in attempting to characterize accurately waters having a substantial organic component and by the fact that, in many cases, a clear-cut separation between waters dominated by mineral and organic acids may not exist. We presume that the effect of acid deposition on naturally acid, colored waters would be to increase the mineral acid component and, thus, the loading of ionic aluminum species. If the loading exceeds the capacity of the organic component to complex the aluminum in solution, there would be a tendency to switch from a relatively nontoxic colored organic system to a clear system dominated by mineral acid anions and ionic aluminum species. This switch could occur as a result of both protonation of organic acids and an overloading of the complexing capacity for aluminum, but we would expect the latter process to be most important.

7.4 Sensitivity to Water Acidification

The task of predicting the sensitivity of waters to acidification has proven extraordinarily difficult. Various models have been proposed (e.g., Henriksen

1979, 1980; Thompson 1982), but these tend to be subject to severe limitations resulting from differences among watersheds that affect both the capability to export alkalinity to surface waters and the change in alkalinity levels that may occur as a result of any particular level of acid deposition. We hope that the concepts presented herein will be useful in defining the effect of soil properties on sensitivity to water acidification.

Although some biological effects, such as a decreased diversity of phyto-plankton, may occur in surface waters before all alkalinity has been exhausted (Almer et al. 1978; Hendrey 1980), the most dramatic effects of alkalinity loss will occur at or near the point of zero alkalinity (Fig. 7.1). According to the concepts presented here, the parameters that control alkalinity in the soil water are base saturation, CO_2 partial pressure, the solubility of the mineral phase controlling aluminum in solution K_{Al}, the thermodynamic bonding energy of the ion exchanger represented as K_s, and the concentration of anions (other than HCO_3^-) in solution.

Alkalinity may be further modified by processes that occur within the lakes or streams themselves. For example, SO_4^{2-} or NO_3^- reduction in the anoxic hypolimmion or sediments will increase alkalinity, as will the uptake of SO_4^{2-}, NO_3^-, or PO_4^{3-} by organisms such as plankton, algae, fungi, or bacteria. Conversely, alkalinity would be reduced by the mineralization and subsequent oxidation of organic nitrogen and sulfur compounds to form NO_3^- and SO_4^{2-}.

Our focus here is on the soil-mediated processes, but even these comprise a formidable list of parameters that should be defined if we are to predict the effect of acid deposition on the surface waters within a catchment area. Unfortunately, these parameters interact in a manner that precludes defining sensitivity in terms of simple critical levels of the parameters involved. The situation is further complicated by the fact that base saturation is the only soil parameter in the list that is commonly determined as a part of routine soil chemical characterization. Even so, it is possible that by a judicious use of the available information, the data collection required to characterize sensitivity in these terms would be modest.

We suggest that sensitivity be defined in terms of (1) the SO_4^{2-} level that would be required to depress alkalinity to zero (or some other level) at some fixed level of CO_2 and (2) the time required to achieve this SO_4^{2-} level at a fixed level of input. In the event that zero alkalinity would not be achieved when the SO_4^{2-} levels were at equilibrium with this fixed input, the time parameter would be expressed in terms of the estimated time for sufficient cation export to occur to achieve zero alkalinity. The latter would be estimated from exchangeable base reserves, assuming weathering is zero, and would, thus, be a worst-case assumption. A rough approximation of the level at which SO_4^{2-} is likely to stabilize and the time required to achieve this stabilization can be made using the concepts described in Chap. 3. Although the use of a fixed level of CO_2 in defining this sensitivity is arbitrary, certainly a system in which the alkalinity would be negative at about 3% CO_2 would have to be considered endangered. This CO_2 level might be set on a regional rather than national basis if data are sufficient to establish time-integrated regional differences in CO_2 contents. An alternative scheme might be to define sensitivity in terms of the CO_2 level below which the alkalinity would be negative, at a fixed SO_4^{2-} level.

If sensitivity is defined in terms of SO_4^{2-} and CO_2 as suggested herein, and base saturation estimates are considered to be available from existing data bases (or with an acceptable level of data collection for systems of special interest), our data input requirements for a specific system are reduced to obtaining estimates for K_{Al} and K_s. Although the practicality of obtaining estimates of these parameters is currently somewhat speculative, the difficulties do not seem insurmountable. Most soil chemical characterizations contain at least semiquantitative estimates of soil mineralogy, and it is entirely possible that, with a modest amount of laboratory correlation studies, these could be used to estimate K_{Al}. If K_{Al} and base saturation are known, only the lime potential, which can be calculated from the soil pH measured in dilute $CaCl_2$, need be known to estimate K_s (Sect. 5). Having this information for any specified soil CO_2 level, the SO_4^{2-} concentration required to produce zero alkalinity and/or the time required to reduce the base saturation to the point at which a specified level of input would result in zero alkalinity could be estimated.

Aside from the perhaps radically different scheme of assessing water sensitivity suggested herein, it should be understood that there are correlations between current methods and the concepts we have set forth. For instance, the model of Henriksen (1979, 1980) was originally conceived as a titration model in which the amount of SO_4^{2-} increase was considered to be equal to the decrease in alkalinity. In our terms, this would be equivalent to the assumption that $\Delta M/\Delta H$ is equal to zero. Henriksen notes that this assumption will not be valid in the presence of significant "washout" of calcium or magnesium salts from the watershed in our example, values of $\Delta M/\Delta H$ near zero will occur only in systems having very low base saturation. In Fig. 5.7, $\Delta M/\Delta H$ does not fall below 0.5 until base saturation drops below 2%. In later papers, Henriksen (1982) and Wright (1983) introduced a factor F into the Henriksen model to correct for this washout and attempts are made to evaluate this factor. They eventually arrive at an estimate of 0.4 (Henriksen 1982; Wright 1983). To the extent that surface water chemistry is determined by contact with the soil system, this F factor is, in fact, $\Delta M/\Delta H$ (Fig. 5.7). In view of the steep slope at low values of $\Delta M/\Delta H$ it seems unlikely that a general value of this factor can be derived from one system and then extrapolated to other systems or that values of this factor derived from acidified lakes can be applied to nonacidified lakes. A low value of $\Delta M/\Delta H$ would seem to be the most likely reason that a particular lake or group of lakes have become acid when subjected to acid deposition, so that a $\Delta M/\Delta H$ value derived from acidified lakes would be biased towards low values when applied to nonacidified systems.

The importance of the $\Delta M/\Delta H$ concept in determining the sensitivity of surface waters to acidification can hardly be overemphasized. Although it is implicit in the scheme we have suggested herein, it does not appear as a separate and specifically identified parameter. An alternative scheme might use current alkalinity of surface waters along with an estimate of $\Delta M/\Delta H$ to define sensitivity. This method would actually have a good deal in common with the Henriksen model or with the related cation denudation rate (CDR) model of Thompson (1982) but would use the concepts we have set forth here to estimate $\Delta M/\Delta H$ for a particular system.

7.5 Capacity versus Intensity

A good deal of the controversy over the capability of acid deposition to cause water acidification seems to stem from a misunderstanding of the role of capacity and intensity factors in the system. Several authors (Krug and Frink 1983; Sollins et al. 1980; Ulrich 1980; Nilsson et al. 1982) have pointed out that the capacity to generate H^+ by natural processes, such as HCO_3^- leaching, plant uptake, etc., is much greater than that of the anthropogenic inputs. We maintain that the capability to acidify waters passing through acid soil depends not so much on the capacity to furnish H^+ ions as on the intensity with which these are released from the soils to the streams. To the degree that the mobile SO_4^{2-} or NO_3^- ions pass through the soil to the waters, these ions provide a mechanism for increasing the intensity of H^+ in the waters. In fact, the H^+ released may well have been generated and stored in the soil by natural processes. These natural processes may be thought of as forerunners of acid release, which is triggered by anthropogenic, strong-acid mineral anions. In this context, the very processes that are often correctly cited as evidence that natural acidification processes have a much greater capacity for soil acidification than do anthropogenic acid inputs are, in fact, the processes that increase the sensitivity of the system to water acidification triggered by these anthropogenic inputs.

In retrospect, it seems that one of the major problems has been an overreliance on a cation export model of acidification. For example, Galloway et al. (1983) have proposed a model of water (and watershed) acidification in seven stages. Although we agree with many of the concepts presented by these authors, their scheme appears to rely too heavily on the idea that water acidification follows as a result of soil acidification resulting from cation export. Questions then arise as to the capacity of the loading to cause the necessary degree of acidification; calculations suggest this would require a minimum of several decades even in sensitive systems. Although such export may occur, many systems when subjected to acid loading may not require significant cation export to trigger water acidification, and, thus, the question of insufficient capacity to acidify no longer arises. Seip (1980) took a similar approach in proposing a mechanism for acidification of surface waters in Norway. He suggested, quite correctly, that introduction of a mobile anion to an acid soil could immediately result in mobilization of H^+ and aluminum without further soil acidification via cation depletion.

An important characteristic of a mechanism that does not require cation export is its reversibility. The salt-effect/CO_2-degassing mechanism we propose will be reversible to the extent that solution SO_4^{2-} concentrations are reversible, and replenishing the exchangeable cations through the slow process of mineral weathering is unnecessary. Obviously, the cation export process and the salt effect process are not mutually exclusive, and significant cation export as a result of acid loading may occur either before or after the switch from positive to negative alkalinity in solution. To the extent that such cation export occurs, the reversibility will be reduced.

8. Soil Sensitivity

The subject of the sensitivity of soils to acid deposition loading has generally been addressed in the context of sensitivity to base depletion by cation leaching. Undoubtedly, this depletion is a major concern and needs to be addressed. However, there are other probable effects, including mobilization of ionic aluminum species, that may be of equal or, in some cases, greater concern in terms of their probable effect on ecosystems. In this chapter, we shall attempt to consider these processes in the context of the principles and the model we have discussed in the previous chapters. The principles are, of course, closely related to the subject of the sensitivity to surface waters, which we discussed in Chap. 7 on aquatic effects. Although it is not our objective to develop a new or revised scheme of sensitivity classification, we hope that the material in this chapter will be useful to those who might undertake such a task in the future.

8.1 Base Cation Depletion

McFee (1980) defines four parameters as important in estimating soil sensitivity to acid precipitation:

1. The total buffering capacity or cation exchange capacity (CEC) provided primarily by the clay and soil organic matter

2. The base saturation of that exchange capacity, which can be estimated by the pH of the soil

3. The management system imposed on the soil—whether it is cultivated and amended with fertilizers or lime or is renewed by flooding or other additions

4. The presence or absence of carbonates in the soil profile.

These criteria generally address the question of leaching of exchangeable base cations, and for the most part provide a reasonable basis for estimating base cation loss as a result of acid loading. The estimation of base saturation by soil pH is questionable because pH depends on several other parameters as well. Solution anion concentrations, CO_2 partial pressures, the solubility of aluminum in the system, K_{Al}, and the aluminum–calcium selectivity coefficient K_s all affect pH even in the relatively simple model we have used, but nonetheless, there will be a general relationship between pH and base saturation.

Several authors have pointed out that $\Delta M/\Delta H$ decreases as base saturation decreases, so that base cation leaching will decrease at low base saturation. The inference, then, is that the maximum sensitivity to acidification will be found in soils having moderate or moderately high base saturation but a very low CEC (Wiklander 1980; Johnson 1981). This is also implied by a statement by McFee (1980) that "if the soil has a low CEC and a circumneutral pH, then acid inputs are likely to reduce pH rapidly." The implication is that the base saturation will be reduced more by acid deposition in these soils than in soils in which base saturation is low and, therefore, $\Delta M/\Delta H$ is much less than 1.0. We believe it is appropriate to distinguish between sensitivity to acidification as measured by depletion of exchangeable base cations and as measured by changes in pH of the soil solution, a distinction not generally made by previous authors.

We find it useful to examine the sensitivity relationships using the model described in Chap. 5. The $\Delta M/\Delta H$ relationship (see Fig. 5.7) is particularly useful in this regard, and we have chosen to modify that figure here by adding the effect of changing levels of CO_2 (Fig. 8.1). It is apparent from Fig. 8.1 that, in this example at least, the reduction in cation export associated with decreasing base saturation does not become an important factor except at very low base-saturation levels. Over most of the range $\Delta M/\Delta H$ remains relatively constant at a value that is dependent on the CO_2 partial pressure. These values are about 0.96, 0.90, and 0.80 for 0.03, 1, and 5% CO_2, respectively. Except at very low base-saturation levels, most of the SO_4^{2-} input serves to reduce base saturation by leaching base cations. At all CO_2 levels the value of $\Delta M/\Delta H$ does not drop below 0.7 until calcium saturation is <5% of the exchange sites, and values <0.5 generally do not occur until calcium saturation is <2%. The exact relationship depends on the values of K_{Al} and K_s, so the illustration cannot be taken as representative of all soils, but very low values of K_s are required before $\Delta M/\Delta H$ reductions will occur at significantly higher base saturation levels. Such low values are most likely uncommon, although Wiklander (1980) reported substantial depression in $\Delta M/\Delta H$ at 60% base saturation using a pure bentonite clay.

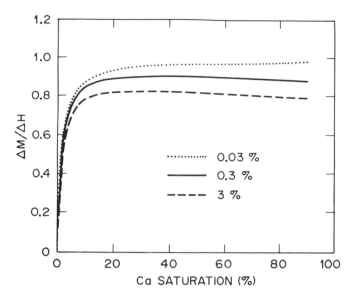

Figure 8.1. Predicted $\Delta M/\Delta H$ as a function of calcium saturation at three CO_2 levels.

Although values of $\Delta M/\Delta H$ substantially <1.0 probably occur almost exclu-
sively at very low base saturation, we would not conclude that this situation is
necessarily rare in acid-affected forest ecosystems. Forest soils often have low
base saturation, and if the natural base saturation level is not sufficiently low to
cause low values of $\Delta M/\Delta H$, the amount of cation export required to bring
about this condition may be small, particularly in soils having low CEC.

On the whole, we tend to agree with previous authors (Wiklander 1980;
McFee 1980; Johnson 1981) that because $\Delta M/\Delta H$ values substantially <1.0 are
usually encountered at low base saturation, maximum sensitivity to base depletion will be found in soils with moderate to moderately high base saturation and
low CEC. However, this agreement does not extend to sensitivity to acidification as measured by pH changes, as will be discussed in Sect. 8.2. We must also
recognize that although low values of $\Delta M/\Delta H$ are associated with low cation
export, we cannot conclude that such soils are immune from deleterious effects
(e.g., Krug and Frink 1983). On the contrary, such soils are sensitive to aluminum mobilization effects, which are discussed more fully in Sect. 8.3.

One implication of the fact that $\Delta M/\Delta H$ will be fairly constant and near 1.0
over most of the base-saturation range, is that, in this range, reduction of base
saturation per unit of H^+ input will be nearly a linear function of CEC and the
total amount of bases that must be removed before soil solutions are no longer
dominated by base cations is roughly the product of the base saturation times
the CEC. Therefore, for reduction of base cations, we would agree with McFee
(1980) that the major parameters affecting sensitivity are CEC and base saturation, at least in unamended soils that do not contain free carbonates.

One extremely important sensitivity criteria for base cation depletion is the rate of soil weathering. Unfortunately, this factor can be only qualitatively assessed in most cases (e.g., type of primary mineral). Quantitative estimates of soil weathering are rare and often based on questionable assumptions (e.g., that net cation export equals weathering rate, which inherently assumes that exchangeable bases remain constant). The reduction of base saturation due to cation export could be significantly modified by varying weathering rates. Weathering is often totally ignored and thus, by default, assumed to be zero. This is clearly not valid and points to the need for more quantitative information on soil weathering.

Base cation loading in atmospheric inputs will also affect the base cation status of soils. Net accumulation of one or more base cations is possible even in cases in which total atmospheric acid deposition loadings are very high. For example, as noted in Chap. 6, Ca^{2+}, Mg^{2+}, and K^+ are accumulating at the Solling site in West Germany even though net loading exceeds 1.5 keq of H^+ per year (Ulrich 1980). The loss or gain of divalent base cations can be predicted by comparing the lime potential log K_L (Chap. 5) of the incoming solution with that of the soil solution. If, after concentration by evapotranspiration processes, the lime potential is greater than that of the soil solution, there will be a net accumulation of divalent cations. Although we have previously considered the divalent cations in terms of Ca^{2+} as an approximation of $Ca^{2+} + Mg^{2+}$, accumulation of the individual cations is important from the standpoint of forest nutrition, so we must remember that log K_L is actually given by pH $-$ 1/2 $p(Ca^{2+} + Mg^{2+})$. The difficulty is that the relevant K_L is not that of the incoming precipitation but of the precipitation after concentration by evapotranspiration processes, as lime potential changes with concentration. Because the degree of concentration that occurs will vary with the rainfall pattern and evapotranspirative demands, it is possible that a particular system may accumulate divalent base cations at times while losing them at other times. As the degree of concentration increases there will be an increasing tendency to leach divalent base cations.

8.2 Sensitivity to pH Changes

Logically, the development of criteria defining the sensitivity of soils to acidification resulting from acid inputs requires defining what measures of acidification will be applied. Most sensitivity schemes focus on the cation export process, but changes in H^+ concentration, whether expressed directly in concentration units or in pH, are not in direct proportion to changes in calcium saturation (see Figs. 5.4 and 5.5).

A reasonable estimate of the effect of acid inputs on acidification might be defined as $d(pH)/d(H)$ (i.e., the change in pH per unit of H^+ input). The plot of pH versus base saturation shown in Fig. 5.5 can be thought of as a plot of pH versus M, where M is the amount of base cation, expressed in units of the CEC of the particular soil. The slope of this curve is therefore $d(pH)/d(M)$, which when multiplied by $d(M)/d(H)$ (or $\Delta M/\Delta H$) from Fig. 5.7 or 8.1 gives us $d(pH)/d(H)$.

The shape of $d(pH)/d(H)$ curve (Fig. 8.2) is very interesting and probably would not have been anticipated by most workers, including the authors. The units of the dependent variable are $d(pH)$ per H^+ input, where the H^+ input is expressed in units of CEC. Thus, the values of less than 1.0 encountered over the range of 20 to 80% calcium saturation mean that an input of H^+ equal to 1% of CEC would cause a change of <0.01 pH units. As calcium saturation decreases below 20%, the pH change per unit of H^+ input increases, reaching 3 (or 0.03 pH units for an H^+ input equal to 1% of CEC) at about 2.6% calcium saturation. As calcium saturation declines further, the pH change per unit of H^+ input declines precipitously. Mathematically, the peak at low calcium saturation arises because $\Delta M/\Delta H$ does not decline as rapidly as does the slope of the pH versus calcium saturation curve, resulting in an increase in the product of the two. Conceptually, the peak may be thought of as lying between the aluminum and the ion exchange buffer ranges (Ulrich 1983). As divalent base cations are depleted from the exchange system, those remaining are more tightly held and, thus, are less available for pH buffering of the solution. When this effect is expressed at calcium saturation and pH levels above that at which the buffering resulting from mineral dissolution is strongly expressed, a peak in the $d(pH)/d(H)$ curve will result.

The most important implication of the shape of this curve is that the sensitivity to pH change is actually quite different from the sensitivity to base cation leaching. Maximum sensitivity to pH change would not result from the combination of moderate-to-high base saturation and low CEC, as implied by several

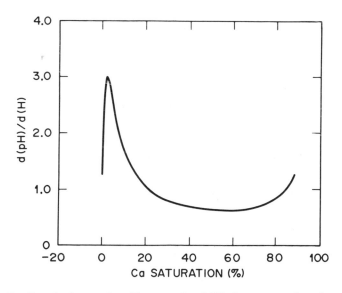

Figure 8.2. Predicted change in pH per unit of H^+ input as a function of calcium saturation.

workers (e.g., Wiklander 1980; McFee 1980; Johnson 1981), but rather a combination of low CEC and low base saturation. Although we have not investigated
the effect of changing the values of K_{Al} and K_s on the shape of the curve, we
assume it is sensitive to these parameters. This curve must, at this time, be
considered only a testable hypothesis because it rests on theoretical rather than
experimental results. However, we would point out that the assumption that
pH would change more rapidly per unit of H^+ input at high base saturation than
at low base saturation rest largely on a much less rigorous theoretical analysis.

8.3 Sensitivity to Aluminum Mobilization

The handling of sensitivity to aluminum mobilization in sensitivity classification schemes has been highly variable. Of the three authors mentioned above,
Wiklander (1980) makes no mention of aluminum mobilization as a possible
result of acid deposition, McFee (1980) notes the problem in his discussion but
does not directly address it in the proposed scheme, and Johnson (1981) does
specifically include criteria for sensitivity to aluminum mobilization. An increase in SO_4^{2-} concentration will result in a corresponding increase in cations
in solution, part of which will be base cations and the remainder, H^+ and
aluminum species. The fraction that is base cations is given by $\Delta M/\Delta H$,
whereas the fraction of H^+ and aluminum species is $1 - (\Delta M/\Delta H)$. Reductions
in sensitivity to base cation loss that are brought about by low base saturation
and the accompanying low values of $\Delta M/\Delta H$, will, therefore, be accompanied
by increased levels of solution aluminum (i.e., aluminum mobilization). Decreases in sensitivity to base cation loss brought about by reduced values of
$\Delta M/\Delta H$ are obtained only at the price of increased aluminum mobilization.

To illustrate further the aluminum mobilization effect, we have reproduced
the aluminum curves from Fig. 5.4b and d on a single graph (Fig. 8.3). As
before, the 25 μeq SO_4^{2-} line represents the normal case and 250 μeq SO_4^{2-}
represents the situation under moderate-to-heavy acid deposition loading. As
SO_4^{2-} concentrations increase as a result of acid deposition, the effect is to
change from the 25- to the 250-μeq SO_4^{2-} line, concurrently increasing the
amount of cations in solution. At moderate-to-high levels of base saturation, in
the example above perhaps 20% calcium saturation, the increase in aluminum
in solution will be small and probably will have no practical effect. In this base-
saturation range, aluminum mobilization resulting from acid impact would not
be significant until cation export occurred to reduce base saturation. Aluminum
mobilization would, thus, depend on cation export, and sensitivity to aluminum
mobilization would depend on base saturation and CEC in the same manner as
does base cation loss.

At low base saturation, however, there is a substantial spread between the
two lines, so that significant aluminum could be brought into solution simply by
the concentration increase associated with impact. Sensitivity to aluminum
mobilized in this way is independent of CEC and, for a particular set of K_{Al} and
K_S values, depends only on base saturation. The role of the aluminum solubility

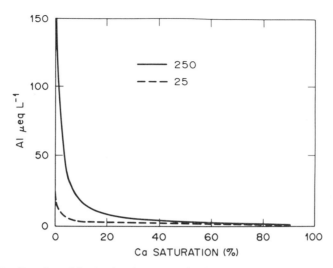

Figure 8.3. Predicted total ionic aluminum species for total solution concentration of 25 and 250 μeq/L. See Chap. 5 for details of input parameters.

parameter K_{Al} and the selectivity coefficient K_S may well prove to be crucial in defining sensitivity to aluminum mobilization. Unfortunately, a good deal of further investigation will probably be required before these parameters can be effectively be included in classification schemes.

8.4 Summary

In summary, we are concerned with three major effects in soils (i.e., cation depletion, aluminum mobilization, and pH depression). On a broad scale, the soil parameters that determine the degree to which each of these effects will occur are the CEC and the base saturation. In the case of base cation depletion, soil weathering must also be considered. Although the precise base-saturation levels at which effects will be observed varies among soils depending on K_{Al} (i.e., the solubility of aluminum in the system,) and on the bonding energies of the exchange sites, K_S. The values of the constants are not usually known, so the sensitivity evaluation must be based simply on base saturation and CEC. Because a general relationship exists between base saturation and pH in soils, soil pH is sometimes used as an approximation of base saturation in sensitivity classification.

The total reservoir of exchangeable base cations is the product of the base saturation times CEC. Soils having high base saturation and high CEC will be insensitive to cation depletion whereas soils having moderate base saturation and low CEC will be depleted much more rapidly. Base depletion, however, may not occur in soils having very low base saturation, regardless of CEC, and in these soils, aluminum and H^+ will be leached rather than exchangeable bases.

Aluminum mobilization effects are only likely to be observed at very low base saturation. Therefore, soils having very low base saturation will be sensitive to aluminum mobilization because of acid deposition, even though CEC may be high. However, in soils that are not naturally very low in base saturation, the rate of reduction of base saturation through cation depletion depends on the CEC. Soils having low CEC will reach the base-saturation levels at which aluminum mobilization occurs much more rapidly than soils having high CEC.

The sensitivity of pH depression to acid deposition is more difficult to define in terms of base saturation and CEC than is sensitivity to cation depletion or aluminum mobilization. We suggest the maximum pH depression per unit of H^+ input will occur at the base saturation at which the transition from cation depletion effects to aluminum mobilization effects is taking place.

With one important exception, the conventional view that increasing CEC decreases sensitivity appears to be substantially correct for any of the three effects with which we are concerned (i.e., base cation loss, pH depression, or aluminum mobilization). For a given base-saturation level, the amount of H^+ input required to effect a given change in any of these parameters varies directly with CEC. Soils having high base saturation and high CEC will be insensitive to changes induced in any of these three parameters by acid deposition, and although soils having high CEC and low base saturation are not immune to these effects, the amount of input required per unit of change will be comparatively high. The important exception to this rule is that if base saturation is sufficiently low that increased solution concentrations will mobilize aluminum, this effect will be independent of CEC.

The effect of base saturation on sensitivity is more complex. In general, soils having moderate-to-high base saturation will lose more base cations than soils having very low base saturation, although in most soils, significant reduction in cation export may require very low levels of base saturation. High-base-saturation soils will be only moderately sensitive to pH change, but there may be a peak in pH sensitivity near the low end of the base-saturation scale. Low base saturation soils will tend to be sensitive to aluminum mobilization (organic soils where the major exchangeable cation is likely to be H^+ rather than Al^{3+} may be a special case), but much depends on whether or not significant aluminum will be mobilized without further base cation loss. In some low-base-saturation soils, significant aluminum mobilization may occur simply as a result of the increased solution concentration resulting from acid deposition, and in this case, sensitivity to aluminum mobilization will not depend on CEC. However, if base saturation is sufficiently high that cation depletion must occur before aluminum will be mobilized, sensitivity to aluminum mobilization will depend on CEC.

9. Conclusion

In Chap. 1, we presented an overview of the processes that occur when acid deposition impinges on a terrestrial ecosystem, and from time to time in our more detailed discussions, we have attempted to summarize again material that would help the reader to understand the significance of the individual processes in terms of the total system. Although a summary at this point is to some degree repetitive, we feel it important to reexamine the total system in the context of the concepts we have developed in the previous chapters. In addition, we are including a final section that points out how our perspective of the system may differ from previously accepted concepts.

9.1 Summary

Acid deposition may be thought of as deposition in which a substantial excess of strong acid anions predominates over basic cations. The major strong-acid anion excess is usually in the form of SO_4^{2-}, although significant amounts of NO_3^- may be deposited and NH_4^+ in the deposition may be converted to NO_3^- by ecosystem processes. In cases in which the influence of marine air masses is strong, substantial Cl^- may be present, although this is usually accompanied by an equivalent amount of basic cations, largely Na^+ and, therefore, does not contribute to the strong acid anion excess. Besides Na^+, basic cations commonly found in deposition include Ca^{2+}, Mg^{2+}, and K^+. However, by definition, acid deposition will contain a significant excess of acid anions over basic cations.

In most naturally acid forest ecosystems, the predominant mobile anions are HCO_3^- and organic acid anions formed by biological processes within the ecosystem. The HCO_3^- is formed by the reaction of CO_2 with water; thus, the HCO_3^- concentration in the soil at any particular time depends on the amount of biological activity such as root and microbial respiration in the soil. As a result, HCO_3^- tends to be highly variable over time. The formation of HCO_3^- from CO_2 and water is depressed by acidity, so that in very acid soils, significant amounts of HCO_3^- may not be present or, at most, will only be present during short periods of high biological activity. Under these conditions, the predominant mobile anions in natural systems are organic acids.

From our perspective, the fundamental perturbation that occurs in acid soils as a result of acid deposition is an increase in solution concentration stemming from the increased supply of strong acid mobile anions, particularly SO_4^{2-}. Charge balance considerations dictate that the increased anion concentration must be accompanied by an equivalent increase in cations in solution, and the nature of the soil chemical processes involved is such that it is accompanied by a shift in the relative proportions of the various cations as well as an increase in the total concentration. This shift will be in the direction of an increase in the proportions of cations having higher valence at the expense of a decreased proportion of cations having lower valence. Specifically, the proportion of the Al^{3+} ion will increase, whereas the proportion of the H^+ ion will actually decrease. This decrease in H^+ is only relative; the absolute concentration of H^+ will actually increase (i.e., there will be a slight decrease in pH). Thus, two general types of effects may be expected (1) an increase in the rate of base cation removal as a result of acid deposition and (2) an increase in the concentration of H^+ in the solution accompanied by an increase in concentration of ionic aluminum species, particularly Al^{3+}. These effects are not mutually exclusive, but the importance of each may be vastly different in different systems. The degree to which either of these effects may occur and the likelihood of consequences to either the terrestrial ecosystems or downstream aquatic systems is highly variable, depending to a large degree on the chemical properties of the system (many of which are not well understood) and on the tolerance of the biological components of the system.

Acid precipitation entering a forest ecosystem first encounters the canopy, an encounter that may result in marked changes in chemical composition. Increased acidity may result from either a washout of organic acids, such as may occur in conifers, especially those growing on low-base-status soils or from the washout of acid dry deposition from the canopy. This later effect is very strong at the Solling site in West Germany (Ulrich 1980) and in the pine barrens of New Jersey (Turner et al. 1985). Decreased acidity may occur as a result of an exchange of H^+ ions in the precipitation for base cations in leaves or other plant tissues in the canopy, a process that is particularly prevalent in high-base-status deciduous stands during the growing season.

The washout of organic acids is a natural process. From our current perspective, we need only be aware that it can occur and that the simple observation of low pH in throughfall does not necessarily indicate that the system is impacted by acid deposition. Increased acidity resulting from washout of dry deposition does, however, represent a net increase in acid loading over and above the amount in rainfall. The decreased acidity that may be observed as a result of ion exchange in the canopy is more interesting. To the extent that this process returns base cations prematurely from the canopy to the forest floor, it could conceivably cause nutrient deficiencies or imbalances in the canopy, but generalization in this regard is hazardous at best. An important point concerning this ion exchange relates to the "neutralization" of incoming acids. When such exchange occurs, the pH of throughfall may be higher than that of the incoming rainfall. This neutralization, however, does not represent a lessening of the capacity of the acid deposition to acidify the system. That this should be true on a macroscale is obvious from a simple consideration of the input–output fluxes of base cations. If we consider the total soil–plant system to the bottom of the rooting zone, an output of base cations leached in association with the incoming strong acid mobile anions must necessarily result in acidification of the system, regardless of internal flows.

It is also useful to view this canopy-leaching process from a different perspective. In the natural system, the bases will return to the forest floor either as leachates in the form of salts of weak organic acids or as bicarbonates on decomposition of the organic material. Either form contributes to the alkalinity of the forest floor (Chap. 7). If, under acid loading these bases are returned as neutral salts of strong acids, they no longer contribute to the alkalinity and the net result is a more acid system.

When acid deposition impinges upon a previously unaffected soil, the net result is an increased SO_4^{2-} concentration. Over the long term, we can expect that a new SO_4^{2-} concentration will be established. This equilibrium concentration will be somewhat higher than that in the incoming precipitation because of the effect of dry deposition and the concentration resulting from evapotranspiration. The increased SO_4^{2-} concentration is subject to two important delay mechanisms (i.e., biological uptake and SO_4^{2-} retention in the soil). Although the delay resulting from biological immobilization may be substantial, in general, we consider the soil retention mechanisms to be the more important, both because of the magnitude of the effect and the variability among soils. In soils having a low retention capacity, the sulfate front resulting from a change in impact may move downward through a 1-m depth within a few weeks or even less. In soils having a high adsorbing capacity, the passage of this front may require a decade or more per meter.

The SO_4^{2-} retention capacity of most soils is concentration dependent, resulting in three important effects. First, because the capacity for adsorption of SO_4^{2-} depends on the solution concentration, it is also dependent on the rate of acid loading. Therefore, the size of the reservoir of adsorbed SO_4^{2-} will change with changes in the rate of loading. Second, the rate at which the front moves through the soil usually is not affected much by the rate of SO_4^{2-} loading. This is particularly important in the interpretation of research results in which artificially high rates of loading have been used to accelerate effects. In fact, level of loading affects the equilibrium concentration above the front but is not likely to affect significantly the rate of migration of the front. Third, if adsorption effects cause a substantial delay in elevated SO_4^{2-} concentrations at the bottom of the root zone after the onset of increased loading, a similar delay may be encountered in reduction of SO_4^{2-} concentrations after loading is reduced. The nature of this delay depends on the reversibility of the retention process. It may involve both desorption from charged surfaces and the dissolution of hydroxy aluminum SO_4^{2-} minerals. Although this reversibility is known to vary among soils, in general it is not well understood. There is some evidence that the highly podzolized soils prevalent in the northeastern United States and in Scandinavia generally do not adsorb large amounts of SO_4^{2-}, whereas the Ultisols of the southeastern United States are likely to be strong adsorbers. Because the major chemical effects of acid deposition on soils are the consequences of increased solution SO_4^{2-} concentration, the time delay of this increased concentration engendered by SO_4^{2-} retention will also be reflected as a time delay of these effects.

As mentioned above, the most important chemical effects that can be expected as a consequence of increased SO_4^{2-} concentrations are (1) increased levels of base cations in solution, resulting in acceleration of base cation depletion and (2) increased concentrations of aluminum ion species, particularly Al^{3+}, and, to a lesser degree, increased H^+. The conditions that determine the relative importance of these processes in a given situation can be understood by referring to Fig. 8.3. At moderate-to-high base saturation, in this example above 25%, the soil solutions will be dominated by base cations and solution aluminum levels will be low even in the presence of increased SO_4^{2-} levels resulting from acid deposition. Under these conditions, immediate ecosystem effects are likely to be minor, but base cation depletion as a result of the increased base cation content of the drainage water will occur. As base cation export proceeds, the base saturation will decrease, although this depletion may be at least partially balanced by the release of base cations by weathering of soil minerals. Given sufficient time under acid loading, base saturation may be deleted to the point that the balance of cations in solution is substantially altered (i.e., the proportion of base cations decreases and the aluminum species increase). The elevated aluminum concentrations are toxic to many plants and, thus, may alter the character of the ecosystem. Perhaps even more important, they contribute to decreased alkalinity in the drainage waters and negative effects in downstream ecosystems. The rate at which this base cation depletion occurs depends on the size of the reservoir of exchangeable bases in the soil (i.e., the product of the base saturation times the CEC), the rate of replenishment by weathering processes, and the level of deposition. The cation export process is slow, and relatively simple calculations suggest that most soils in affected areas would require many decades or even centuries before deleterious effects could be expected. Because these calculations generally do not take into account the replenishment of base cations by weathering of soil minerals such calculations are likely to underestimate the time required for cation depletion.

Probably more important are the immediate effects of a shift from base cation dominance in soil solutions to the aluminum dominance that may be expected in some low-base-status soils. In our example (Fig. 8.3), at 25 μeq of SO_4^{2-}/L (simulating the normal system), there is a total of only 14 μeq/L of ionic species of aluminum at the very low calcium saturation level of 1%. However, at a solution SO_4^{2-} level of 250 μeq/L (simulating the acid loaded system) 14 μeq of aluminum/L are found at 13% calcium saturation and, at 1% calcium saturation, there would be 130 μeq of ionic aluminum species in solution. In acid soils, the increase in aluminum that occurs as a result of increased solution concentration will be largely in the form of Al^{3+} and the increase will occur in both relative and in absolute amounts. The actual location of the curves shown in Fig. 8.3 would vary with the aluminum–calcium selectivity coefficient and the solubility of the controlling mineral phase, so the example must not be taken as describing any particular system. Yet, the principle seems clear. At low base saturations, the switch from calcium-dominated solutions to solutions containing high levels of aluminum species may occur directly as a result of acid deposition loading without base depletion through cation export. The time lag for this effect to be observed after the onset of acid loading would be largely that associated with SO_4^{2-} retention, and unless significant cation depletion occurred during deposition, the solutions would again shift to base cation dominance if solution SO_4^{2-} levels decreased as a result of reduced loading.

The same principle applies to acidification of surface water, just as the shift to increased levels of aluminum and H^+ in the soil solutions is reflected in a decrease in alkalinity when these solutions are released as surface waters. In the normal condition, the mobile anions consist largely of HCO_3^- and organic acid anions (which will be discussed subsequently). The HCO_3^- levels depend on the acidity of the solution and the partial pressure of CO_2 in the soil. Increasing CO_2 results in the formation of H^+ and HCO_3^-. In the soil, the H^+ formed in this manner reacts with soil minerals to form aluminum species, which, in turn, displace base cations from the exchange complex. The result is that, even in quite acid soils, the solutions may be dominated by HCO_3^- salts of the base cations. In HCO_3^- waters of this type not in contact with soil chemical processes, the pH is highly dependent on the CO_2 partial pressure. On release from the soil to the lower CO_2 environment of the surface waters, the pH will rise substantially. The net result is that acid soils may actually release alkaline waters to the surface. This alkalinity will vary over time as soil CO_2 levels vary. In low base-saturation soils, the increased solution SO_4^{2-} concentration associated with acid loading will result in some increase in H^+, which in turn will decrease HCO_3^-. The aluminum ion species will also increase.

If the increase in aluminum and H^+ is sufficient that the total charge associated with these species exceeds that associated with HCO_3^-, the solution will have a negative alkalinity (i.e., net strong acid acidity) and the pH rise associated with decreased CO_2 levels will be drastically reduced. Such waters remain acid when released from the soil to the surface-water system. Thus, the same processes that result in increased ionic aluminum in soil solutions as a result of acid deposition can result in a shift from alkaline to acid surface waters, even though only a minor change in the pH of the soil solution is observed. In some cases, this shift may not occur until base cations are depleted by decades or even centuries of acid loading, whereas in other cases, the shift may be observed as soon as the soil solution SO_4^{2-} levels increase in response to acid deposition loading.

In many forest soils, organic acids comprise a substantial fraction of the total mobile anions. Organic acid anions are likely to be particularly high in the organic horizons of forest soils, and leachates collected in these horizons of forest soils are characteristically highly colored and may be very acid, especially in coniferous forests growing on low-base-status soils. When these acid solutions enter soil horizons high in iron and/or aluminum minerals, iron and aluminum tend to come into solution, forming complexes with the organic anions. The downward migration of these complexes, leaving silica behind, results in the characteristic "ashy" E horizon found in Spodosols. When the complexing capacity of the organic anions is exceeded, precipitates composed of organocomplexes of iron and aluminum form, resulting in the typical dark colored Bs or illuvial horizons. Waters collected below this point in the profile are likely to be clear and not particularly acid. If waters high in organic acids are discharged directly into surface waters, as may occur in peat bogs or many boreal systems, the result is a naturally acid but usually highly colored water. In some regions, waters of this type are fairly common. For the most part they seem to be less toxic to aquatic life than waters of similar acidity resulting from an input of strong acid mineral anions, apparently because in these organic systems the concentration of the ionic aluminum species is reduced by the capacity of the organic anions to form aluminum complexes.

Nitrogen, either as NO_3^- or NH_4^+, which can be converted to nitrate by ecosystem processes, can be a significant component of acid deposition. Nitrate is a highly mobile anion, and the effect of high levels of NO_3^- in solution on base cation depletion, aluminum mobilization, and water acidification is very similar to the effect of SO_4^{2-}, except that NO_3^- is only very weakly adsorbed by soil systems, so that there is no NO_3^- analog to the SO_4^{2-} adsorption system. However, because the biological demand for nitrogen in forest ecosystems is usually about 15 times the demand for sulfur, this factor is much more important in determining the fate of nitrogen. In general, if nitrogen inputs do not exceed the biological demand for this element, there will be little or no tendency for nitrogen inputs resulting from acid deposition to accelerate base cation depletion or aluminum mobilization. Under these conditions the major effect likely to occur is a fertilization effect, resulting in accelerated growth. This is the most probable effect in cases in which forest growth tends to be nitrogen-limited, as is the case in most of the United States and much of Scandinavia. However, to the extent that nitrogen inputs exceed biological demand, the most likely result is increased NO_3^- levels in solution and increased leaching loss. The effect of this increase in the mobile NO_3^- concentration of the on base depletion and aluminum mobilization will be similar to that of increased SO_4^{2-} concentration. There is some indication that in central Europe and parts of southern Scandinavia nitrogen inputs are sufficiently high that biological demand is exceeded and that NO_3^- is contributing to aluminum mobilization and cation depletion. One difference we might expect between NO_3^- and SO_4^{2-} effects, however, relates to the fact that most nitrogen in the ecosystem will be in organic form and that NO_3^- release will be mediated by biological processes. We can, therefore, expect greater time variation or a "pulse" pattern for high NO_3^- concentrations.

9.2 Concepts in Transition

In conclusion, it seems appropriate to point out some of the differences between the concepts that we have set forth in this book and what seems to have been a generally accepted set of concepts concerning the effect of acid deposition on soils and waters. We realize that considerable differences exist among the concepts put forward by various workers and that some of the concepts that we have presented have been advanced previously, at least in part. Nonetheless, we feel the differences are important.

The importance of the SO_4^{2-} retention properties of the soil in determining the lag time between the onset of acid deposition loading and the appearance of increased SO_4^{2-} concentrations in soil solutions and surface waters has now become reasonably well accepted (Galloway et al. 1983). Here, we do not take issue with the concept but simply wish to emphasize its importance.

Perhaps the major difference between our concepts and those of others lies in our perception of the importance of the mechanism of increased aluminum mobilization in acid soils and reduced (or even negative) alkalinity in surface waters resulting from the increased mobile anion concentrations associated with acid deposition. This "salt effect" mechanism, previously proposed by Seip (1980), when expanded and interpreted in the light of CO_2 partial pressure effects, can account for observed changes in water acidity without the necessity for invoking further soil acidification through base cation loss. Failure to understand this mechanism completely and, thereby, to appreciate its importance, has caused most workers, including the ourselves, to overemphasize the role of the cation depletion mechanism in evaluating probable effects of acid deposition. Although there is no doubt that cation depletion can occur as a result of acid deposition, in our opinion, many observed effects can probably be accounted for simply by the chemical changes associated with increased concentrations of mobile anions.

In terms of the likelihood of deleterious effects, there seems to be both positive and negative implications in a switch from a cation depletion model to one relying principally on the effects of the concentration of mobile anions. In low-base-saturation soils in which negative effects have been delayed by SO_4^{2-} adsorption, we can expect that these effects will be expressed immediately if the solution SO_4^{2-} levels reach the critical point for that system: there will be no further grace period while acidification proceeds via base cation leaching. Conversely, this model seems to imply a somewhat brighter recovery picture than does the cation depletion model. When the concentration of mobile anions diminishes after a reduction in loading, high levels of either solution Al^{3+} or of water acidity probably will not be maintained. Unfortunately, the delay in reduction of mobile anion concentration may be very slow in soils in which large amounts of SO_4^{2-} have been retained.

References

Abrahamsen, G. 1980. "Impact of Atmospheric Sulfur Deposition on Forest Ecosystems, pp. 357–371 in D. S. Shriner et al., eds., *Atmospheric Sulfur Deposition: Environmental Impact and Health Effects*, Ann Arbor Science, Ann Arbor, Mich.

Adams, F., and Hajek, B. F. 1978. "Effects of Solution Sulfate, Hydroxide, and Potassium Concentrations on the Crystallization of Alunite, Basaluminite, and Gibbsite from Dilute Aluminum solutions," *Soil Sci.* 126:169–173.

Adams, F., and Rawajfih 1977. "Basaluminite and Alunite: A Possible Cause of Sulfate Retention by Acid Soils," *Soil Sci. Soc. Am. Proc.* 41:686–691.

Alban, D. H. 1982. "Effects of Nutrient Accumulation by Aspen, Spruce, and Pine on Soil Properties," *Soil Sci. Soc. Amer. J.* 46:853–860.

Almer, B., Dickson, W., Ekstrom, C., and Hornshrom, E. 1978. pp. 271–311 in J. O. Niagru, ed., *Sulfur in the Environment, Part II. Ecological Impacts*, Wiley, New York.

Bache, B. 1974. "Soluble Aluminum and Calcium Aluminum Exchange in Relation to the pH of Dilute Calcium Chloride Suspensions of Acid Soil," *J. Soil Sci.* 25:321–335.

Bettany, J. R., Stewart, W. B., and Saggar, S. 1979. "The Nature and Forms of Sulfur in Organic Matter Fractions of Soils Selected Along an Environmental Gradient, *Soil Sci. Soc. Amer. J.* 43:981–985.

Bornemisza, E., and Llanos, R. 1967. "Sulfate Movement, Adsorption and Desorption in Three Costa Rican Soils," *Soil Soc. Amer. Proc.* 31:356–360.

van Breemen, N. 1973. "Dissolved Aluminum in Acid Sulfate Soils and Acid Mine Waters," *Soil Sci. Soc. Am. Proc.* 37:694–697.

van Breemen, N., Burrough, P. A., Velthorst, E. J., van Dobben, H. F. de Witt, T., Ridder, T. B., Reignders, H. F. R. 1982. "Soil Acidification from Atmospheric Ammonium Sulfate in Forest Canopy Throughfall," *Nature* 229:548–550.

van Breemen, N., and Jordens, E. R. 1983. "Effects of Atmospheric Amonium Sulfate on Calcareous and Non-calcareous Soils of Woodlands in the Netherlands. pp. 171–182 in B. Ulrich and J. Pankrath, eds., *Effects of Accumulation of Air Pollutants on Forest Ecosystems*, Reidel, Boston.

Chao, T. T., Harward, M. E., and Fang, S. C. 1962a. "Adsorption and Desorption Phenomena of Sulfate Ions in Soils," *Soil Sci. Soc. Amer. Proc.* 26:234–237.

Chao, T. T., Harward, M. E., and Fang, S. C. 1962b. "Soil Constituents and Properties in the Adsorption of the Sulfate Ion," *Soil Sci.* 94:286–293.

Chao, T. T., Harward, M. E., and Fang, S. C. 1965. "Exchange Reactions Between Hydroxyl and Sulfate Ions in Soils," *Soil Sci.* 99:104–107.

Chen, C. W., Gherini, S., Hudson, R. J. M., and Dean, J. D. 1983. *The Integrated Lake-Watershed Acidification Study. Vol. 1: Model Principles and Procedures*, EA-3221, vol. 1. Electric Power Research Institute. Palo Alto, Calif.

Christopherson, N., Seip, H. M., and Wright, R. F. 1982. "A Model for Streamwater Chemistry at Birkenes, Norway," *Water Resour. Res.* 18:977–996.

Clark, S. S., and Hill, R. G. 1964. "The pH-Percent Base Saturation Relationship of Soils," *Soil Sci. Am. Proc.* 28:490–492.

Cole, D. W., and Rapp, M. 1981. "Elemental Cycling in Forest Ecosystems", pp. 341–409 in D. E. Reichle, ed., *Dynamic Properties of Forest Ecosystems,* Cambridge University Press, London.

Coleman, N. T., and Thomas, G. W. 1967. "The Basic Chemistry of Soil Acidity," pp. 1–34 in R. W. Pearson and F. Adams, eds., *Soil Acidity and Liming,* American Society of Agronomy Madison, Wis.

Coulter, B. S., and Talibuddeen, O. 1968. "Calcium-Aluminum Exchange Equilibria in Clay Minerals and Acid Soils," *J. Soil Sci.* 19:237–250.

van den Driessche, R. 1971. "Response of Conifer Seedlings to Nitrate and Ammonium Sources of Nitrogen," *For. Sci.* 18:126–132.

Fitzgerald, J. W., Strickland, T. C. and Swank, W. T. 1982. "Metabolic Fate of Inorganic Sulphate is Soil Samples from Undisturbed and Managed Forest Ecosystems," *Soil Biol. Biochem.* 14:529–536.

Gaines, G. L., and Thomas, H. C. 1953. "Adsorption Studies on Clay Minerals: A Formulation of the Thermodynamics of Exchange Adsorption," *J. Chem. Phys.* 21:714–718.

Galloway, J. N., Norton, S. A., and Church, M. R. 1983. "Freshwater Acidification from Atmospheric Deposition of Sulfuric Acid: A Conceptual Model," *Environ. Sci. Technol.* 17:541A–545A.

Grier, C. C. 1975. "Wildfire Effects on Nutrient Distribution and Leaching in a Coniferous Ecosystem," *Can. J. For. Res.* 5:595–607.

Harward, M. E., and Reisenauer, H. M. 1966. "Reactions and Movement of Inorganic Sulfur," *Soil Sci.* 101:225–236.

Henderson, G. S., Harris, W. F., Todd, D. E., and Grizzard, T. 1977. "Quantity and Chemistry of Throughfall as Influenced by Forest Type and Season," *J. Ecol.* 65:364–374.

Hendrey, G. R. 1980. "The Effects of Acidity on Primary Productivity in Lakes: Phytoplankton," pp. 357–371 in D. S. Shriner et al., eds., *Atmospheric Sulfur Deposition: Environmental Impact and Health Effects,* Ann Arbor Science, Ann Arbor, Mich.

Henriksen, A. 1979. "A Simple Approach to Identifying and Measuring Acidification in Fresh Water," *Nature* 278:542–544.

Henriksen, A. 1980. "Acidification of Fresh Waters—a Large Scale Titration," pp. 68–74 in D. Drablos and A. Tollan, eds., *Ecological Effects of Acid Precipitation,* Johs. Grefslie Trykkeri A/S, Mysen, Norway.

Henriksen, A. 1982. *Changes in Base Cation Concentrations Due to Acid Precipitation,* OF-81623, Norwegian Institute for Water Research, Oslo, Norway.

Höfken, K. D. 1983. "Input of Acidifiers and Heavy Metals to a German Forest Area Due to Wet and Dry Deposition," pp. 57–64 in B. Ulrich and J. Pankrath, eds., *Effects of Accumulation of Air Pollutants on Forest Ecosystems,* Reidel, Boston.

Johnson, D. W. 1985. Sulfur cycling in forests. Biogeochemistry 1:29–43.

Johnson, D. W. 1981. "Effects of Acid Precipitation on Elemental Transport from Terrestrial to Aquatic Ecosystems," pp. 539–545 in R. A. Fazzolare and C. Smith, eds., *Beyond the Energy Crisis: Opportunity and Challenge,* Pergamon Press, New York.

Johnson, D. W., and Cole, D. W. 1980. "Anion Mobility in Soils: Relevance to Nutrient Transport from Terrestrial Ecosystems," *Environ. Int.* 3:79–90.

Johnson, D. W., Cole, D. W., Gessel, S. P., Singer, M. J., and Minden, R. W. 1977. "Carbonic Acid Leaching in Tropical, Temperate, Sub-alpine, and Northern Forest Soil," Arct. Alp. Res. 9:329–343.

Johnson, D. W., Henderson, G. S., Huff, D. D., Lindberg, S. E. Richter, D. D.,

Shriner, D. S., and Turner, J. 1982a. "Cycling of Organic and Inorganic Sulphur in a Chestnut Oak Forest," *Oecologia* 54:141–148.

Johnson, D. W., Richter, D. D., Lovett, G. M., and Lindberg, S. E. 1985. "Effects of Acid Deposition on Cation Nutrient Cycling in Two Deciduous Forests, *Can. J. For. Res.* 15:772–782.

Johnson, D. W., and Todd, D. E. 1983. "Some Relationships Among Fe, Al, C, and SO_4^{2-} in a Variety of Forest Soils," *Soil Sci. Soc. Amer. J.* 47:792–800.

Johnson, D. W., West, D. C., Todd, D. E., and Mann, L. K. 1982b. "Effects of Saw-log vs. Whole-tree Harvesting on the Nitrogen, Phosphorus, Potassium, and Calcium Budgets of an Upland Mixed Oak Forest," *Soil. Sci. Soc. Am. J.* 56:1304–1309.

Kononova, M. M. 1966. *Soil Organic Matter,* Pergamon Press, New York.

Krug, E. C., and Frink, C. R. 1983. "Acid Rain on Acid Soil: A New Perspective," *Science* 221:520–525.

Lee, J. J., and Weber, D. E. 1982. "Effects of Sulphuric Acid Irrigation on Major Cation and Sulphate Concentrations on Water Percolating Through Two Model Hardwood Forests, *J. Environ. Qual.* 11:57–64.

Lindberg, S. E., Harriss, R. C., Turner, R. R., Shriner, D. S., and Huff, D. D. 1979. Mechanisms and rates of atmospheric deposition of selected trace elements and sulfate to a deciduous forest watershed. ORNL/TM-6674. Oak Ridge National Laboratory, Oak Ridge, Tennessee. 514 p.

Lindsay, W. L. 1979. Chemical Equilibria in Soils. John Wiley and Sons, New York.

Matzner, E. 1983. "Balance of Element Fluxes Within Different Ecosystems Impacted by Acid Rain," pp. 147–155 in B. Ulrich and J. Pankrath, eds., *Effects of Accumulation of Air Pollutants on Forest Ecosystems,* Reidel, Boston.

Matzner, E., and Ulrich, B. 1983. "The Turnover of Protons by Mineralization and Ion Uptake," pp. 93–103 in B. Ulrich and J. Pankrath, eds., *Effects of Accumulation of Air Pollutants on Forest Ecosystems,* Reidel, Boston.

McCormick, L. H., and Steiner, K. C. 1978. "Variation in Aluminum Tolerance Among Six Genera of Forest Trees," *For. Sci.* 24:565–568.

McFee, W. W. 1980. "Sensitivity of Soil Regions to Long Term Acid Precipitation," pp. 495–506 in D. S. Shriner et al., ed., *Atmospheric Sulfur Deposition: Environmental Impact and Health Effects,* Ann Arbor Science, Ann Arbor, Mich.

Meiwes, K. J., and Khanna, P. K. 1981. "Distribution and Cycling of Sulphur in the Vegetation of Two Forest Ecosystems in an Acid Rain Environment," *Plant Soil* 60:369–375.

van Miegroet, H., and Cole, D. W. 1982. Potential Effect of Acid Rain in Soil Nutrient Status and Solution Acidity—Controlling Mechanisms, in *Proceedings, 1982 West Coast Regional Meeting, Portland, Oregon,* National Council of the Pulp and Paper Industry for Air and Stream Improvement, Inc. (in press).

van Miegroet, H. and Cole, D. W. 1984. "The Impact of Nitrification on Soil Acidification and Cation Leaching in a Red Alder Ecosystem," *J. Environ. Qual.* 13:586–590.

Miller, H. 1983. "Studies of Proton Fluxes in Forests and Heaths in Scotland," pp. 183–193 in B. Ulrich and J. Pankrath, eds., *Effects of Accumulation of Air Pollutants on Forest Ecosystems,* Reidel, Boston.

Miller, H. G., Cooper, J. M., Miller, J. D., and Pauline, O. J. L. 1979. Nutrient Cycling in Pine and Their Adaptation to Poor Soils," *Can. J. For. Res.* 9:19–26.

Mollitor, A. V. and Raynal, D. J. 1982. Acid Precipitation and Ionic Movements in Adirondack Forest Soils," *Soil Sci. Amer. J.* 46:137–141.

Nilsson, S. I., and Bergkvist, B. 1983. "Aluminum Chemistry and Acidification Processes in a Shallow Podzol on the Swedish West Coast," *Water Air Soil Pollut.* 20:311–329.

Nilsson, S. I., Miller, H. G. and Miller, J. D. 1982. "Forest Growth as a Possible Cause of Soil and Water Acidification: An Examination of the Concepts," *Oikos* 39:40–49.

Nordstrom, D. K. 1982. The Effect of Sulfate on Aluminum Concentrations in Natural

Waters: Some Stability Relations in the System Al_2, O_3, $-SO_4$, $-H_2O$ at 298 K°. *Geochim. Cosmochim. Acta* 46:681–692.

Persson, G. 1982. *Acidification Today and Tomorrow,* Swedish Ministry of Agriculture, Environment, '82 Committee, Translated by Simon Harper.

Peterson, L. 1976. Podzols and podzolization. DSR Forlag, Copenhagen.

Pleysier, J. L., Juo, A. S. R. and Herbillon, J. 1979. "Ion Exchange Equilibria Involving Aluminum in a Kaolinitic Ultisol," *Soil Sci. Soc. Am. J.* 43:875–880.

Posner, A. M. 1964. *Titration Curves of Humic Acid,* pp. 161–174 in *Eighth International Congress of Soil Science, Bucharest, Romania. Vol. II.*

Prenzel, J. 1983. "A mechanism for Storage and Retrieval of Acid in Acid Soils," pp. 157–170 in B. Ulrich and J. Pankrath, eds., *Effects of Accumulation of Air Pollutants on Forest Ecosystems,* Reidel, Boston.

Rajan, S. S. S. 1979. "Adsorption and Desorption of Sulfate and Charge Relationships in Allophanic Clays, *Soil Sci. Amer. J.* 43:65–69.

Rehfuess, K. E., Bosch, C., and Pfannkuch, E. 1982. "Nutrient Imbalances in Coniferous Stands in Southern Germany, presented at International Workshop on Growth Disturbances of Forest Trees. IUFRO/FFRJ-Jyvasyla, Finland, October 10–13, 1982.

Reuss, J. O. 1975. *Chemical/biological Relationships Relevant to Ecological Effects of Acid Rainfall,* EPA-660/3-75-032, Corvallis, Oreg.

Reuss, J. O. 1977. "Chemical and Biological Relationships Relevant to the Effect of Acid Rainfall on the Soil-Plant System," *Water Air Soil Pollut.* 7:461–478.

Reuss, J. O. 1978. *Simulation of Nutrient Loss from Soils Due to Rainfall Acidity,* EPA-660/3-78-053, Corvallis, Oreg.

Reuss, J. O. 1980. "Simulation of Soil Nutrient Losses Resulting from Rainfall Acidity," *Eco. Mod.* 11:15–38.

Reuss, J. O. 1983. "Implications of the Ca-Al Exchange System for the Effect of Acid Precipitation on Soils," *J. Environ. Qual.* 12:591–595.

Reuss, J. O. and Johnson, D. W. 1985. Effect of Soil Processes on the Acidification of Water by Acid Deposition, *J. Environ. Qual.* 14:26–31.

Richter, D. D., Johnson, D. W., and Todd, D. E. 1983. "Atmospheric Sulfur Deposition, Neutralization, and Ion Leaching in Two Deciduous Forest Ecosystems," *J. Environ. Qual.* 12:263–270.

Schindler, D. W. 1980. "Implications of Regional Scale Lake Acidification," pp. 534–538 in D. S. Shriner et al., ed., *Atmospheric Sulfur Deposition: Environmental Impact and Health Effects,* Ann Arbor Science, Ann Arbor, Mich.

Schnitzer, M., and Khan, S. U. 1972. *Humic Substances in the Environment,* Marcel Dekker, New York.

Schofield, C. W. 1980. "Processes Limiting Fish Populations in Acidified Lakes," pp. 346–356, in D. S. Shriner et al., ed., *Atmospheric Sulfur Deposition: Environmental Impact and Health Effects.* Ann Arbor Science, Ann Arbor, Mich.

Schofield, R. K., and Taylor, A. W. 1955. "Measurements of Activities of Bases in Soils," *J. Soil Sci.* 6:137–146.

Seip, H. M. 1980. "Acidification of Fresh Waters: Sources and Mechanisms," pp. 358–566, in D. Drablos and A. Tollan, ed., *Ecological Effects of Acid Precipitation,* Johs. Grefslie Trykkeri A/S, Mysen, Norway.

Singh, B. R., Abrahamsen, G., and Stuanes, A. 1980. "Effect of Simulated Acid Rain on Sulfate Movement in Acid Forest Soils," *Soil Sci. Soc. Amer. J.* 44:75–80.

Singh, S. S., and Brydon, J. E. 1970. Activity of Aluminum Hydroxy Sulfate and the Stability of Hydroxy Aluminum Interlayers in Montmorillonite, *Can. J. Soil Sci.* 50:219–225.

Skeffington, R. A. 1983. "Soil Properties Under Tree Species in Southern England in Relation to Acid Deposition in Throughfall," pp. 219–231 in B. Ulrich and J. Pankrath, eds., *Effects of Accumulation of Air Pollutants on Forest Ecosystems,* Reidel, Boston.

Smith, R. A., and Alexander, R. B. 1983. *Evidence for Acid Precipitation-induced Trends in Stream Chemistry at Hydrologic Bench-mark Stations,"* U.S. Geological Survey Circular 910. Alexandria, Va.

Sollins, P. C., Grier, C. C., Corison, F. M., Cromack, K., Fogel, R., and Fredrickson, R. L. 1980. "The Internal Element Cycles of an Old-Growth Douglas-fir Ecosystem in Western Oregon," *Ecol. Mon.* 50:261–285.

Stednick, J. D. 1982. "Sulfur Cycling in Douglas-fir on a Glacial Outwash Terrace," *J. Environ. Qual.* 11:43–45.

Thompson, M. E. 1982. "The Cation Denudation Rate as a Quantitative Index of Sensitivity of Eastern Canadian Rivers to Acidic Atmospheric Precipitation," *Water Air Soil Pollut.* 18:215–226.

Turner, J., and Lambert, M. J. 1980. Sulfur Nutrition of Forests, pp. 321–344 in D. S. Shriner et al., ed., *Atmospheric Sulfur Deposition: Environmental Impact and Health Effects,* Ann Arbor Science, Ann Arbor, Mich.

Turner, R. C., and Clark, S. S. 1964. Lime Potential and Degree of Base Saturation of Soils. *Soil Sci.* 99:194–199.

Turner, R. S., Johnson, A. H., and Wang, D. C. 1985. *Biogeochemistry of Aluminum in McDonald's Branch Watershed, New Jersey Pine Barrens, J. Environ. Qual.* 14:314–323.

Ugolini, F. C., Minden, R., Dawson, H., and Zachara, J. 1977. An example of soil processes in the Abies amabilis zone of Central Cascades, Washington. *Soil Sci.* 124:291–302.

Ulrich, B. 1980. *Production and Consumption of Hydrogen Ions in the Ecosphere,* pp. 255–282 in T. C. Hutchinson and M. Havas, ed., *Effects of Acid Precipitation on Terrestial Ecosystems,* Plenum, New York.

Ulrich, B. 1983. "Soil Acidity and Its Relations to Acid Deposition," pp. 127–146, in B. Ulrich and J. Pankrath, eds., *Effects of Accumulation of Air Pollutants on Forest Ecosystems,* Reidel, Boston.

Ulrich, B., Mayer, R. and Khanna, P. K. 1980. "Chemical Changes due to Acid Deposition in a Loess-derived Soil in Central Europe, *Soil Sci.* 130:193–199.

Wiklander, L. 1980. "The Sensitivity of Soils to Acid Precipitation, pp. 553–568 in T. C. Hutchinson and M. Havas, ed., *Effects of Acid Precipitation on Terrestrial Ecosystems,* Plenum, New York.

Wiklander, L., and Andersson, A. 1972. "The Replacing Efficiency of Hydrogen Ion in Relation to Base Saturation and pH," *Geoderma* 7:159–165.

Wood, T., and Bormann, F. H. 1975. "Increases in Foliar Leaching Caused by Acidification of an Artificial Mist," *Ambio* 4:169–171.

Wright, R. F. 1983. *Predicting Acidification of North American Lakes,* Report 4/1983, Norwegian Institute for Water Research, Oslo, Norway.

Appendix: Model Documentation

The model documentation consists of four sections. Section A.1, "Model Description," gives the theoretical basis of the model, the derivation of the equations that are used in the program code, and the methods used to obtain the solutions to these equations. The "Program Operation" section (A.2) contains step by step instructions for the use of the BASIC program. The "Program Documentation" section (A.3) gives a complete description of the program code, which is listed in Section A.4.

A.1 Model Description

The following ionic species are considered in the model:

Input	Unknown
SO_4^{2-}	H^+
$NO_3^- + Cl^-$	Ca^{2+}
	Al^{3+}
	$Al(OH)^{2+}$
	$Al(OH)^+$
	HCO_3^-

The model consists of a set of six simultaneous equations. When values are supplied for the concentrations of SO_4^{2-} and $NO_3^- + Cl^-$ in the soil solution and for four additional input parameters, the solution of these equations gives the soil-solution concentrations of the six "unknown" ion species.

Additional Input Parameters:

CaX: The fraction of the ion-exchange sites occupied by calcium.

P_{CO_2}: The partial pressure of CO_2 in the system.

$\log K_A$: The value of the log of the equilibrium constant $K_A°$ [Eq. (A-4)], that is the value of 3 pH $-$ pAl appropriate to the system.

$\log K_S$: The Gaines-Thomas aluminum–calcium selection coefficient (Gaines and Thomas 1953).

The chemical equations describing the formation of the three aluminum species (i.e., Al^{3+}, $Al(OH)^{2+}$, and $Al(OH)_2^+$) are

$$Al(OH)_3 + 3\ H^+ \rightleftarrows Al^{3+} + 3\ H_2O, \tag{A-1}$$

$$Al^{3+} + H_2O \rightleftarrows Al(OH)^{2+} + H^+, \tag{A-2}$$

$$Al^{3+} + 2\ H_2O \rightleftarrows Al(OH)_2^+ + 2\ H^+. \tag{A-3}$$

The $\log K°$ value for Eq. (A-1) varies among soils and is, therefore, treated as an input parameter. The $\log K°$ value used for Eq. (A-2) is -5.02, and for Eq. (A-3) it is -9.30 (Lindsay 1979). The equilibrium expressions for Eqs. (A-1), (A-2), and (A-3) are given by Eqs. (A-4), (A-5), and (A-6), respectively:

$$\frac{\gamma_3[Al^{3+}]}{\gamma_1^3[H^+]^3} = K_A, \tag{A-4}$$

$$\frac{\gamma_2\gamma_1[Al(OH)^{2+}][H^+]}{\gamma_3[Al^{3+}]} = 10^{-5.02}, \tag{A-5}$$

$$\frac{\gamma_1\gamma_2[Al(OH)_2^+][H^+]^2}{\gamma_3[Al^{3+}]} = 10^{-9.30}. \tag{A-6}$$

The brackets denote concentrations (mol/L), while γ_1, γ_2, and γ_3 are the activity coefficients for the mono-, di-, and trivalent ions, respectively.

The aluminum–calcium exchange relationship used is that of Gaines and Thomas (1953):

$$\frac{\gamma_3^2[Al^{3+}]^2}{\gamma_2^3[Ca^{2+}]^3} \cdot K_s = \frac{(AlX)^2}{(CaX)^3}. \tag{A-7}$$

Here, K_s is the selection coefficient, and AlX and CaX are the equivalent fractions of exchangeable aluminum and calcium, respectively. CaX is obtained by dividing the equivalents of exchangeable calcium (per unit of soil) by the cation-exchange capacity (CEC). In this version of the model, we are dealing with a two-component exchange system, so that

$$AlX = 1 - CaX. \tag{A-8}$$

Substituting for AlX in Eq. (A-7) we obtain (A-9):

$$\frac{\gamma_3^2[Al^{3+}]^2}{\gamma_2^3[Ca^{2+}]^3} \cdot K_s = \frac{(1 - CaX)^2}{(CaX)^3}. \tag{A-9}$$

The reaction of CO_2 and water to form HCO_3^- is

$$CO_2 + H_2O \rightleftarrows H^+ + HCO_3^-. \tag{A-10}$$

The log $K°$ value is -7.81 (Lindsay 1979), and the equilibrium expression is

$$\frac{\gamma_1^2[H^+][HCO_3^-]}{P_{CO_2}} = 10^{-7.81}. \tag{A-11}$$

where P_{CO_2} is the partial pressure of CO_2 in atmospheres.

Finally, the model utilizes the charge-balance concept [Eq. (A-12)], that is, the positive and negative charges in soil solution must be equal. The set of the six equations that actually comprise the model is shown in Eqs. (A-12) through (A-17). In Eqs. (A-13) through (A-17), the concentration of each of the other unknown ion species is expressed in terms of $[H^+]$. Equation (A-13) is simply a rearrangement of Eq. (A-4).

$$3 [Al^{3+}] + 2 [Ca^{2+}] + 2 [Al(OH)^{2+}] + [Al(OH)_2^+] + [H^+]$$

$$= [HCO_3^-] + 2 [SO_4^{2-}] + [NO_3^- + Cl^-], \tag{A-12}$$

$$[Al^{3+}] = \frac{K_{Al}\gamma_1^3[H^+]^3}{\gamma_3}, \tag{A-13}$$

$$[Al(OH)^{2+}] = \frac{(\gamma_1^2 K_{Al} \cdot 10^{-5.02})[H^+]^2}{\gamma_2}, \tag{A-14}$$

$$[Al(OH)_2^+] = (K_{Al} \cdot 10^{-9.30})[H^+], \tag{A-15}$$

$$[Ca^{2+}] = \frac{\gamma_3^2 K_s(CaX^3)^{2/3}}{\gamma_2^3(1 - CaX)^2} \cdot \frac{\gamma_1^3 K_A^{°1/3}}{\gamma_3} \cdot [H^+]^2, \tag{A-16}$$

$$[HCO_3^-] = \frac{P_{CO_2} \cdot 10^{-7.81}}{\gamma_1^2} \cdot \frac{1}{[H^+]}. \tag{A-17}$$

Eqs. (A-14) and (A-15) are obtained by substituting the right side of Eq. (A-13) for $[Al^{3+}]$ in Eqs. (A-5) and (A-6), respectively, followed by rearrangement. Similarly, Eq. (A-16) is obtained by substituting the right side of (A-13) into (A-9) and solving for $[Ca^{2+}]$. Eq. (A-16) is not shown in its simplest form, because the form shown is more convenient for the use of temporary variables in the solution. Eq. (A-17) is a simple rearrangement of A-11.

In solving this set of equations (A-12 through A-17), it was convenient to define a series of temporary variables, A, B, C, D, and K_c:

$$A = \frac{\gamma_1^3 K_{Al}}{\gamma^3}, \tag{A-18}$$

$$B = \frac{\gamma_1^2 K_{Al} \cdot 10^{-5.02}}{\gamma_2}, \tag{A-19}$$

$$C = K_{Al} \cdot 10^{-9.30}, \tag{A-20}$$

$$D = \frac{\gamma_3^2 K_S (CaX)^3}{\gamma_2^3 (1 - CaX)^2}, \tag{A-21}$$

$$K_c = \frac{P_{CO_2} \cdot 10^{-7.81}}{\gamma_1^2}. \tag{A-22}$$

Eqs. (A-13) through (A-17) can now be rewritten as Eqs. (A-23) through (A-27), respectively:

$$[Al^{3+}] = A \, [H^+]^3, \tag{A-23}$$

$$[Al(OH)^{2+}] = B \, [H^+]^2, \tag{A-24}$$

$$[Al(OH)^+] = C \, [H^+], \tag{A-25}$$

$$[Ca^{2+}] = A^{1/3} D^{2/3} \, [H^+]^2, \tag{A-26}$$

$$[HCO_3^-] = \frac{K_c}{[H^+]}. \tag{A-27}$$

Substituting the above into the charge-balance equation (A-12) and moving all nonzero terms to the left-hand side, we obtain

$$3A[H^+]^3 + 2A^{1/3}D^{2/3} \, [H^+]^2 + 2B[H^+]^2 + [H^+]$$

$$- \frac{K_c}{[H^+]} - 2 \, [SO_4^{2-}] - [NO_3^- + Cl^-] = 0.$$

Multiplying through by $[H^+]$ and collecting terms,

$$3A[H^+]^4 + 2(A^{1/3}D^{2/3} + B) \, [H^+]^3 + (1 + C) \, [H^+]^2$$
$$-(2 \, [SO_4^{2-}] + [NO_3^- + Cl^-]) \, [H^+] - K_c = 0. \tag{A-28}$$

To facilitate the programming, one further temporary variable, A_D, is defined such that

$$A_D = A^{1/3} D^{2/3} + B, \tag{A-29}$$

and Eq. (A-28) is rewritten as

$$3A[H^+]^4 + 2A_D \, [H^+]^3 + (1 + C) \, [H^+]^2 - (2 \, [SO_4^{2-}]$$
$$+ [NO_3^+ + Cl^-]) \, [H^+] - K_c = 0. \tag{A-30}$$

Equation (A-30) is of the form

$$f([H^+]) = 0.$$

Only the real nonzero solution is of interest, and it occurs at the value of $[H^+]$ such that $f([H^+])$ is equal to zero. The value of $f([H^+])$ is calculated for successive trial values of $[H^+]$ until convergence, that is, until the calculated value of $f([H^+])$ approaches zero (within preset limits).

The actual procedure followed to select trial values is best understood by referring to Fig. A.1. For any set of input parameters, a plot of $f([H^+])$ versus $[H^+]$ will assume the general shape shown in Fig. A.1. The program assumes an initial trial value of $[H^+]$ based on pH 6.0 and $f([H^+])$ is calculated. Next, the value of the first derivative, $f'([H^+])$, is calculated using Eq. (A-31):

$$f'([H^+]) = 12A\,[H^+]^3 + 6A_D\,[H^+]^2 + 2(1 + C)\,[H^+]$$
$$-2\,[SO_4^{2-}] - [NO_3^- + Cl^-]. \qquad \text{(A-31)}$$

For any value of $[H^+]$, the slope of the curve is given by $f'([H^+])$. Using this shape, a new trial value of $[H^+]$ is obtained by projecting a tangent from the curve at the point $([H^+], f([H^+]))$ to the horizontal axis, that is, the line $f(H^+) = 0$. This process is then repeated until successive values of $[H^+]$ are within prescribed limits. The method works equally as well if the calculated value of $f([H^+])$ is less than zero, unless the value of $[H^+]$ is such that the slope of the curve is negative. In that case, $f'([H^+])$ evaluates to a negative value and the program automatically inserts a higher trial value of $[H^+]$.

In addition to the input parameters, the calculation of $f([H^+])$ and $f'([H^+])$ requires that the value of the activity coefficients, γ_1, γ_2, and γ_3, be known. Again, initial values are assumed. At each iteration the concentration of each of the individual ions corresponding to the current trial value of $[H^+]$ is calculated

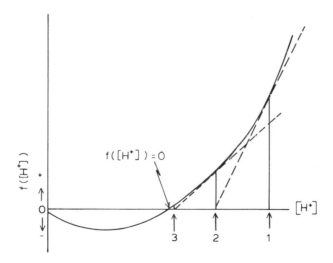

Figure A.1. Schematic diagram showing curve of $f([H^+])$ vs $[H^+]$. The value of $[H^+]$ at which $f([H^+]) = 0$ is obtained by successively calculating $f([H^+])$ for trial values of $[H^+]$ (1), finding the slope by calculating the value of the first derivative, that is, $f'([H^+])$, and projecting tangents to the horizontal axis for new estimates of $[H^+]$, as shown by (2) and (3).

using Eqs. (A-14) through (A-17). The ionic strength μ is then calculated by

$$\mu = 1/2 \; \Sigma \; c_i Z_i^2, \tag{A-32}$$

where μ is the ionic strength, c_i is the concentration of ion i in moles per liter, Z_i is the valence of ion i, and Σ indicates the summation over the products of each of the concentrations multiplied by the corresponding valence. The activity coefficients are then calculated using the Davies form of the Debye-Huckel equations (Lindsay 1979):

$$\log \gamma_i = 0.509 \; Z_i^2 \; \frac{\mu^{1/2}}{1 + \mu^{1/2}} - .3\mu. \tag{A-33}$$

   Although not strictly a part of the model, the lime potential, defined as pH − 1/2 pCa, for any combination of CaX, K_s, and K_{Al}, will also be calculated by the program. The lime potential is independent of solution concentration but varies with CaX. The equation used in the program is

$$pH - 1/2 \; pCa = 1/6 \log \frac{K_s(CaX)^3}{(1 - CaX)^2} + 1/3 \log K_{Al}. \tag{A-34}$$

It is obtained by substituting the right side of Eq. (A-13) for Al^{3+} in Eq. (A-19), followed by rearranging and converting to the logarithmic form.

A.2 Program Operation

The program is menu driven. On startup, the program name and version are displayed along with the prompt PRESS ANY KEY FOR MENU. On pressing any key, the menu is displayed and the user may select from the following options:

   1. INPUT DATA

   2. SINGLE VALUE RUN

   3. BASE SATURATION SERIES

   4. PRINT OUTPUT

   5. OUTPUT TO DISK

   6. END

   On startup, option No. 1, INPUT DATA, must be selected first. Program prompts will then ask the user to supply the following input information (Values shown in parentheses are those used to generate the examples shown in Figs. A.2 and A.3)

   SO4: Enter SO_4^{2-} concentration in μeq/L. (250)
   CL: Enter $(Cl + NO_3^-)$ concentration in μeq/L. (20)

ENTER CO2 PERCENT: Enter percentage CO_2 in the soil. (2)

ENTER LOG GT SELECTIVITY COEFFICIENT: Enter log of the Gaines-Thomas selection coefficient. (2.5)

ENTER LOG KAL: Enter log of the K_{Al} value for Eq. (A-4). (8.5)

The program will not return to the menu, and the user may select either option No. 2 (SINGLE VALUE RUN) or option No. 3 (BASE SATURATION SERIES). If the SINGLE VALUE RUN (option No. 2) is selected, the prompt TURN ON PRINTER Y/N appears. If N is entered, output will only be displayed. Entering a Y will result in both printed and displayed output.

The next prompt displayed is ENTER BASE SATURATION VALUE $.01 < BS < 0.9$. Note that this base saturation (actually Ca + Mg saturation) is entered as a decimal value and not as a percentage. For our example (Fig. A.2), we enter 0.15 (i.e., 15% base saturation). The values of log K_{Al} (KA), the aluminum-calcium selection coefficient (KS), and the lime potential calculated according to Eq. (A-34) are now displayed. Some delay (usually <20 s) may be encountered while the initial solutions are calculated. The following information is then displayed:

BS: base saturation

PH: pH

H+: H^+ (concentration in μeq/L)

CA: Ca^{2+} (concentration in μeq/L)

AL3+: Al^{3+} (concentration in μeq/L)

AL2+: $Al(OH)^{2+}$ (concentration in μeq/L)

AL1+: $Al(OH)_2^+$ (concentration in μeq/L)

HCO3: HCO_3^- (concentration in μeq/L)

CL: Cl^- (concentration in μeq/L)

SO4: SO_4^{2-} (concentration in μeq/L)

ALK: alkalinity   (concentration in μeq/L)

CO_2 (%)

If printed output has been requested by entering a Y at the start of the routine, the results will be both displayed and printed. An example is shown in Fig. A.2.

The next prompt displayed is

ENTER C FOR CHECK ROUTINE

ENTER M FOR MENU

Entering an M returns the program to the menu for another selection. Entering a C activates the CHECK ROUTINE so that the accuracy of the solutions may

```
KAL = 8      KS = 2.5      LIMEPOT = 2.69490599

BS    PH    H+    CA     AL3+   AL2+   AL1+   HCO3   CL    SO4    ALK     CO2%

.15  4.664 22.26 256.42 3.88   1     1.12  14.67  20    250   -13.58  2

*********************************************************************************


KAL = 8.5    KS = 2.5      LIMEPOT = 2.86157266

BS    PH    H+    CA     AL3+   AL2+   AL1+   HCO3   CL    SO4    ALK     CO2%

.15  4.822 15.49 267.5  4.14   1.53  2.45  21.1   20    250   -2.5    2

*********************************************************************************


KAL = 9      KS = 2.5      LIMEPOT = 3.02823932

BS    PH    H+    CA     AL3+   AL2+   AL1+   HCO3   CL    SO4    ALK     CO2%

.15  4.981 10.74 277.6  4.38   2.33  5.38  30.44  20    250   7.6     2

*********************************************************************************


KAL = 9.5    KS = 2.5      LIMEPOT = 3.19490599

BS    PH    H+    CA     AL3+   AL2+   AL1+   HCO3   CL    SO4    ALK     CO2%

.15  5.141 7.44  286.67 4.6    3.53  11.78 44.02  20    250   16.67   2

*********************************************************************************
```

Figure A.2. Sample program output using menu option No. 2, SINGLE VALUE RUN, with optional printed output. Results shown are for log K_{Al} values of 8.0, 8.5, 9.0, and 9.5. Remaining input parameters set as described in Sect. A.2.

be checked. This is particularly useful if alterations have been made in the program; it guards against errors in programming or logic that may otherwise go undetected. In this routine, the program first calculates and displays the total charge (μeq/L) associated with both cations and anions. Next, using concentrations and activity coefficients as calculated by the most recent solution, the program calculates and displays the value of the equilibrium constant K_{Al} [Eq. (A-4)] along with the input value. Similarly, the values of the equilibrium constants for Eqs. (A-5), (A-6), and (A-11) are calculated from the solutions and displayed with the respective values used internally by the program for these constants. Finally, the program calculates and displays both the left- and right-hand sides of the ion exchange equation (A-9). Any discrepancies between these pairs of values in excess of acceptable rounding errors would indicate either an error or that the internal convergence limits used in the program are too large.

Menu option No. 3 calculates solutions for a set of internally generated, base-saturation values. The solutions are stored in the output array and may be printed or stored automatically on a disk after calculations are complete. Alternatively, menu options No. 4 and No. 5 may be used for printing and storage. In this version, the intervals between base-saturation values are reduced in three steps as base-saturation values decrease. This method allows for points to be closely spaced at the low base-saturation levels of most interest for acid-deposition effects, while the execution time is kept to a minimum and the number of rows used in the output is kept below the current DIMENSION value of 100. Ranges and intervals are as follows:

Base-saturation range	Interval
0.8–0.18	0.03
0.175–0.055	0.015
0.050–0.004	0.005

The ranges and intervals for the three steps may be changed by changing the values of the variables IA (upper limit), IB (lower limit), and IC (interval) in lines 105 (first step), 135 (second step), and 145 (third step).

When option No. 3 is selected, the first prompt to appear is ENTER 1 FOR DIRECT TO DISK. If a 1 is entered the next prompt is ENTER NAME OF STORAGE FILE. After calculations are complete, the program will now automatically store the output array on the disk under the file name entered. The next prompt is ENTER 1 FOR DIRECT TO PRINT. If a 1 is entered, the program will provide a printout of the output array without further operator action.

As the calculations proceed, the results for each internally generated base-saturation value are displayed. To review a particular set of results, program execution may be halted by using the "Ctrl-S" command. Program operation will resume when any key is pressed. After calculations are complete and the results have been stored on the disk and printed (if these options were selected), the program returns to the menu.

Menu option No. 4 is PRINT OUTPUT. This option provides a printout showing the current values of the $\log_{10}$ of the aluminum solubility constant, K_{Al}, the $\log_{10}$ of the ion-exchange constant K_s and the current values stored in the output array. The output array includes the base saturation, pH, concentrations of all ions considered in the system (both inputs and those calculated by the program), percentage CO_2, and alkalinity. An example of a printout showing the results of a base-saturation series is shown in Fig. A.3. The printout may contain either the results from a base-saturation series or a series of single-value runs made on a particular set of input data. Selection of either menu option No. 1, INPUT DATA, or No. 3, BASE-SATURATION SERIES, resets the output-array row counter to zero.

Special sets of solutions, such as those generated by the SO4 SERIES or the CO2 SERIES versions of the programs (see Sects. A.3 and A.4), may also be printed using option No. 3.

Menu option No. 5, OUTPUT TO DISK, is used to store the current values in the output array to a disk. When this option is selected, the prompt ENTER NAME OF STORAGE FILE is displayed. The values in the array are then stored under the name entered. The file structure used is compatible with that required by a number of standard plotting or data-management programs. The

LOG K SELECTIVITY = 2.5   LOG KAL = 8.5

CONCENTRATIONS ARE MICRO-EQ/L

BS	PH	H+	CA	AL3+	AL2+	AL1+	HCO3	CL	SO4	ALK	CO2%
.9	5.46	3.57	357.64	.05	.08	.57	91.91	20	250	87.64	2
.87	5.419	3.92	349.04	.07	.1	.62	83.75	20	250	79.04	2
.84	5.386	4.23	342.38	.09	.11	.67	77.48	20	250	72.38	2
.81	5.356	4.53	336.91	.1	.13	.72	72.4	20	250	66.91	2
.78	5.33	4.82	332.25	.13	.15	.76	68.1	20	250	62.25	2
.75	5.305	5.09	328.15	.15	.17	.81	64.36	20	250	58.15	2
.72	5.282	5.37	324.47	.17	.18	.85	61.05	20	250	54.47	2
.69	5.261	5.64	321.11	.2	.2	.89	58.05	20	250	51.11	2
.66	5.24	5.92	317.99	.23	.22	.94	55.31	20	250	47.99	2
.63	5.219	6.21	315.06	.27	.25	.98	52.77	20	250	45.06	2
.6	5.199	6.5	312.28	.31	.27	1.03	50.38	20	250	42.28	2
.57	5.179	6.8	309.6	.35	.3	1.08	48.13	20	250	39.6	2
.54	5.16	7.12	307.02	.4	.32	1.13	45.99	20	250	37.02	2
.51	5.14	7.45	304.49	.46	.35	1.18	43.93	20	250	34.49	2
.48	5.12	7.8	301.99	.53	.39	1.24	41.95	20	250	31.99	2
.45	5.099	8.18	299.51	.61	.43	1.3	40.02	20	250	29.51	2
.42	5.078	8.58	297.02	.71	.47	1.36	38.14	20	250	27.02	2
.39	5.057	9.02	294.5	.82	.52	1.43	36.28	20	250	24.5	2
.36	5.034	9.5	291.92	.96	.58	1.5	34.45	20	250	21.92	2
.33	5.011	10.02	289.25	1.13	.64	1.59	32.63	20	250	19.25	2
.3	4.986	10.62	286.45	1.34	.72	1.68	30.8	20	250	16.45	2
.27	4.959	11.29	283.46	1.61	.81	1.79	28.96	20	250	13.46	2
.24	4.93	12.07	280.22	1.96	.93	1.91	27.09	20	250	10.22	2
.21	4.898	12.99	276.61	2.44	1.07	2.06	25.17	20	250	6.61	2
.175	4.856	14.31	271.71	3.26	1.3	2.27	22.85	20	250	1.71	2
.16	4.836	14.98	269.27	3.75	1.43	2.37	21.81	20	250	-.73	2
.145	4.814	15.75	266.56	4.36	1.58	2.5	20.74	20	250	-3.44	2
.13	4.79	16.64	263.47	5.13	1.76	2.64	19.64	20	250	-6.53	2
.115	4.764	17.67	259.89	6.14	1.99	2.8	18.49	20	250	-10.11	2
.1	4.735	18.9	255.6	7.51	2.28	3	17.29	20	250	-14.4	2
.085	4.702	20.4	250.29	9.44	2.65	3.23	16.02	20	250	-19.71	2
.07	4.664	22.29	243.36	12.31	3.16	3.53	14.66	20	250	-26.64	2
.05	4.6	25.79	229.48	19.08	4.24	4.09	12.67	20	250	-40.52	2
.045	4.581	26.94	224.56	21.74	4.62	4.27	12.13	20	250	-45.44	2
.04	4.561	28.25	218.7	25.06	5.08	4.48	11.56	20	250	-51.3	2
.035	4.538	29.75	211.58	29.29	5.64	4.72	10.98	20	250	-58.42	2
.03	4.513	31.51	202.73	34.8	6.32	4.99	10.37	20	250	-67.27	2
.025	4.485	33.6	191.42	42.2	7.19	5.32	9.72	20	250	-78.58	2
.02	4.454	36.12	176.43	52.46	8.31	5.72	9.04	20	250	-93.57	2
.015	4.418	39.25	155.71	67.33	9.81	6.22	8.33	20	250	-114.292	
.01	4.376	43.22	125.54	90.04	11.91	6.85	7.56	20	250	-144.462	
5E-03	4.327	48.44	78.64	127.02	14.97	7.68	6.75	20	250	-191.362	

Figure A.3. Sample model printout for menu option No. 3, BASE SATURATION SERIES. Input parameters were set as described in Sect. A.2.

first value stored is the current value of the output-array row counter (variable IR). The second value is 12, that is, the number of rows (or fields). Next, the alphanumeric identifiers of the 12-column headings are stored as string variables. The output-array values are then stored by rows, with 12 fields within each row.

No provision is made in the program for reviewing the names of current files on the disk, changing the names of current disk files, or changing to a different disk drive. This may be accomplished from the menu using option No. 6, END. Standard operating system commands may then be used to review files, change disks, etc. The program may be reentered at the menu by a GO TO 35 (the menu routine starts at line 35). If this procedure is followed carefully, the current values of input variables and the output array will be preserved.

Rather simple modifications of the program that provide a series of values for a given input parameter can be very useful. Examples are the SO4 SERIES and CO2 SERIES modifications shown in Sect. A.4 and described in Sect. A.3. In these versions, menu option No. 2 is modified to provide a series of SO_4^{2-} concentrations or CO_2 percentages. When the routine is initiated, the program runs the complete series and stores the results in the output array without further operator intervention. The use of the SO4 SERIES version will be described by way of an example.

In the SO_4 SERIES version, as in the standard version, the DATA INPUT option must first be selected. All values are entered in the usual manner; whatever SO_4^{2-} concentration is entered here will be overridden later. When option No. 2 is selected, the first prompt is again TURN ON PRINTER Y/N. Although this print option will operate normally, it is usually more efficient to print the results from the output array using the print option from the menu. Therefore, an N will normally be entered at this point.

The next prompt is

ENTER BASE SATURATION VALUE

BETWEEN 0.01 and 0.9

Upon entry of this value, the program calculates the solutions for the series of SO_4^{2-} concentrations input via the DATA statement at line 735. Results are displayed as calculated and stored in the output array. After calculations are complete, the program returns to the menu. Results may not be printed using menu option No. 4 (Fig. A.4), or stored in a disk file using option No. 5. The SO_4^{2-} concentrations used in the series may be changed by modifying the DATA statement at line 735.

```
LOG K SELECTIVITY = 2.5      LOG KAL = 8.5

CONCENTRATIONS ARE UEQ/L
```

BS	PH	H+	CA	AL3+	AL2+	AL1+	HCO3	CL	SO4	ALK	CO2%
.15	5.15	7.15	55.31	.37	.32	1.13	44.28	20	0	35.31	2
.15	5.091	8.21	73.24	.57	.42	1.3	38.74	20	25	28.24	2
.15	5.042	9.22	92.63	.81	.53	1.46	34.65	20	50	22.63	2
.15	5	10.16	113.02	1.1	.65	1.61	31.54	20	75	18.02	2
.15	4.965	11.05	134.1	1.43	.77	1.75	29.1	20	100	14.1	2
.15	4.934	11.89	155.67	1.8	.89	1.88	27.13	20	125	10.67	2
.15	4.906	12.68	177.6	2.2	1.01	2.01	25.51	20	150	7.6	2
.15	4.882	13.43	199.81	2.64	1.14	2.13	24.15	20	175	4.81	2
.15	4.86	14.15	222.22	3.11	1.27	2.24	22.99	20	200	2.22	2
.15	4.84	14.83	244.79	3.61	1.4	2.35	21.98	20	225	-.21	2
.15	4.822	15.49	267.5	4.14	1.53	2.45	21.1	20	250	-2.5	2
.15	4.805	16.11	290.3	4.69	1.66	2.55	20.32	20	275	-4.7	2
.15	4.789	16.72	313.2	5.28	1.79	2.65	19.63	20	300	-6.8	2
.15	4.775	17.3	336.16	5.88	1.92	2.74	19.01	20	325	-8.84	2
.15	4.761	17.87	359.17	6.52	2.05	2.83	18.44	20	350	-10.83	2
.15	4.749	18.42	382.23	7.18	2.18	2.92	17.93	20	375	-12.77	2
.15	4.737	18.95	405.34	7.86	2.32	3	17.46	20	400	-14.66	2
.15	4.725	19.46	428.47	8.57	2.45	3.08	17.03	20	425	-16.53	2
.15	4.822	15.49	267.5	4.14	1.53	2.45	21.1	20	250	-2.5	2
.15	4.705	20.45	474.81	10.04	2.71	3.24	16.26	20	475	-20.19	2
.15	4.695	20.93	498.01	10.82	2.85	3.32	15.92	20	500	-21.99	2

Figure A.4. Sample printout from the SO4 SERIES version using menu option No. 4, PRINT OUTPUT.

A.3 Program Documentation

The program as described here and as listed in Sect. A.4 is written in the "Applesoft*" version of the BASIC language for use on Apple* microcomputers or other equipment with compatible versions of the BASIC language. Some minor modifications may be required for use with other versions of BASIC. The program works well, but, after many revisions and modifications, it is not a particularly elegant example of the programming art.

This section consists of a detailed description of the program code. A complete listing of the code is given in Section A.4.

INITIAL routine (Lines 0–49)

Lines

0–21	Greeting
22	Dimension arrays
	A(100,12): Output array
	B(12): Temporary output vector
	B$: Column identifier string
30	Strings D$ and DD$ set to CHR$(4) (Ctrl-D) and CHR$(9) (Ctrl-I), respectively. Used for control of output devices.
35–45	These lines set up the menu. The INPUT, SINGLE VALUE RUN, and BASE SATURATION SERIES routines are accessed directly from the menu using the ON IS GO TO statement, where IS is the input variable that determines which of the line numbers in the set the program selects. Because the PRINT OUTPUT and OUTPUT TO DISK routines are also entered from other points in the program, they are set up as subroutines. These subroutines are accessed indirectly from the menu through the GOSUB statements in lines 40 and 42.

INPUT routine (lines 50–99)

50–51	Constants used at various points in the program are defined as variables T1 (10), T2 (100), T3 (1000), T6 (10^6), and T8 (10^8). CK$ string is set to 1 for use as a flag check.
52–56	In response to the prompts SO4 and CL, the user enters the solution concentrations (μeq/L) or SO_4^{2-} and $Cl^- + NO_3^-$. These are converted to mol/L and stored as variables SA and Cl, respectively. LT is defined as $\log_{10}$ (10).
60	The user enters percentage CO_2 in response to the prompt ENTER CO2 PERCENT. This is converted to atmospheres and stored as variable CO.
65	The $\log_{10}$ of the Gaines-Thomas selection coefficient [Eq. (A-9)] and

* "Applesoft" and "Apple" are registered trade marks of Apple Computer, Inc., Cupertino, CA, USA.

the aluminum solubility constant [Eq. (A-4)] are entered in response to the prompts ENTER LOG GT SELECTIVITY COEFFICIENT and ENTER LOG KAL. These are converted to antilogs and stored as variables KS and KA, respectively.

70   Row counter for output array (IR) set to zero. Temporary values set for activity coefficients G1, G2, and G3, and for pH (PH and TP), and H^+ concentration (H^+).

72–77   Defines print divider NP$, and sets up strings NA$, NB$, NC$, and N$, which are used for output headings. A$ is used for formatting, and U$ stores text for output.

78   Alphanumeric field (column) identifiers are stored in the B$( ) string vector fields B(1) − B(12).

80–85   Defines functions of (H^+). F(H) is from Eq. (A-30), and FP(H) is the first derivative of F(H) [Eq. (A-31)].

90   Return to menu.

BASE SATURATION SERIES routine (lines 100–199)

100   Sets the output-array row counter (IR) to 0.

102–103   If operator enters 1 in response to the ENTER 1 FOR DIRECT TO DISK prompt, the flag string (ID$) is set to 1, and control is transferred to line 250 for entry of storage-file name. Control then returns to line 104.

104   If operator enters a 1 in response to the ENTER 1 FOR DIRECT TO PRINT prompt, the flag string (IP$) is set to 1.

105–150   IN this section an initial base saturation value is set and subsequently decremented by steps. A each step the simultaneous equations are solved (SOLUTIONS subroutine at line 400). The results are displayed and stored in the output array. Over the range of 0.9–0.18 the base saturation decrement is set at 0.015, over the range 0.175–0.055 it is 0.005, and over the range 0.050–0.0005 the decrement is 0.0005. Ranges and decrements for each of the three steps may be changed by changing the values of variables IA (initial value), IB (final value), and IC (decrement), in lines 105, 135, and 145, respectively.

160   If the "DIRECT TO DISK" option has been selected (line 102) flag ID$ will be set to 1 and the disk storage subroutine is entered at line 260. Control then returns to line 165.

165   If the "DIRECT TO PRINT" option has been selected (line 104), flag IP$ will be set to 1, and the print output subroutine (line 200) is called.

170   The disk storage and print flags; ID$ and IP$, respectively; are set to 0, and control return to the menu.

PRINT OUTPUT routine (lines 200–249)

200–220   Turns on printer.

225–226   Prints current values of log $K_A{}^\circ$, and K_S, and column headings.

230–240   The output array contains 12 columns (fields) and IR rows. From left to right these fields are: base saturation, pH, the concentrations (μeq/L) of H^+, Ca^{2+}, Al^{3+}, $Al(OH)^{2+}$, $Al(OH)_2^+$, HCO_3^-, ($Cl^- + NO_3^-$), SO_4^{2-}, alkalinity, and %CO_2. Using a temporary string (B$), a string (L$) consisting of the current value of each of these fields is created and printed for each row.

245–246   Printer is turned off and control returned to the point from which the subroutine was called, i.e. either the menu or the base saturation series routine at line 165.

OUTPUT TO DISK routine

250   Storage file name entered is stored as string F$.

253   If OUTPUT TO DISK routine was entered from the BASE SATURATION SERIES routine, the ID$ flag will be set to 1, and control will be transferred back to that routine after entry of the storage file name.

260–265   Monitor is turned on. Any previous file stored under the currently designated file name will be opened and deleted. The file is opened and prepared for writing.

270   The number of rows in the output array (IR) and the number of fields (columns) in each row (12) are a written to the disk.

272   The 12 alphanumeric column identifiers contained in the B4(12) vector are written.

275–285   The values contained in rows 1-IR of the output array are written to the disk (by rows).

290–299   File is closed, monitor turned off, and control returned to the routine from which the OUTPUT TO DISK routine was called, i.e., either the menu or the BASE SATURATION SERIES routine.

SOLUTIONS subroutine (lines 400–600)

400–405   Row counter (IR) for output array is incremented.

410–415   Defines a series of temporary variables A, B, C, D, KC, and AD [see Eqs. (A-19), (A-20), (A-21), (A-22), and (A-29), respectively].

420   Using the current value of H, calculates functions F(H) and FP(H) storing the values as variables FC and FS, respectively. If FS <0, the slope of F(H) is negative, and as the solution will not converge the trial H is multiplied by a factor of 5, and the program returns to line 410. If FS $\geq$0, the program continues to the next line.

425   The intercept of the tangent with the F(H) axis (vertical axis in Fig. A-1), is calculated, and the current value of H is retained as variable FH.

430   A new value of H is calculated, i.e. the intercept of the tangent with the H axis (horizontal axis in Fig. A.1). A negative value indicates the current trial value is too low, in which case the trial value is reset to 10^{-7}.

440   Convergence check. If the change in the value of H over two succes-

sive iterations is less than the prescribed limit (i.e. $\Delta H/H < H/100$), convergence is satisfactory and control passes to line 500. If convergence is not satisfactory the program continues to the next line (450).

450   Using current values of H, concentrations of $Al(OH)_2^+$ (A1), $Al(OH)^{2+}$ (A2), Al^{3+} (A3), Ca^{2+} (CA), and HCO_3^- (BC) are calculated, as per Eqs. (A-23)–(A-27).

465   The ACTIVITY subroutine (line 600) is called to calculate updated values of the activity coefficients G1, G2, and G3.

470   Program returns to line 410 for next iteration.

OUTPUT FORMATTING routine (lines 500–599)

500   Calculates pH

505–525   Current values of base saturation and pH are rounded and stored as variables B(1) and B(2), respectively. Current H^+, Ca^{2+}, Al^{3+}, $Al(OH)^{2+}$, $Al(OH)_2^+$, HCO_3^-, Cl^-, and SO_4^{2-} are converted to $\mu eq/L$, and stored (in that order) as variables B(3)–B(10). CO_2 partial pressure is converted to percent and stored as variable B(12). Alkalinity is calculated and stored in B(11).

530   Concentration values in variables B(3)–B(11) are rounded.

540–548   A string (L$) is created, containing the output values for base saturation, pH, H^+, and Ca^{2+}. String L$ is displayed under the alphanumeric headings contained in string NA$. The information in string L$ is then transferred to string LL$, and L$ is reset to the null string ("").

550–552   String L$ is re-created containing the values for Al^{3+}, $Al(OH)^{2+}$, $Al(OH)_2^+$, and HCO_3^-, displayed under the headings in string NB$, an added to string LL$. L$ is again set to the null string.

555–560   String L$ is created a third time, using the values for CL^-, SO_4^{2-}, alkalinity, and percentage CO_2, displayed under the headings contained in NC$ and again added to string LL$. Current values have now been stored both as rounded real numbers in the B( ) vector and in alphanumeric form in print format in the LL$ string.

565   The print divider (*****.....) is displayed.

570–575   Current output values are transferred to line IR in the output array. Control is returned to either the BASE SATURATION SERIES routine at line 125 or to the SINGLE VALUE RUN routine at line 760.

ACTIVITY COEFFICIENT routine (lines 600–699)

600–630   This subroutine is called in each iteration through the SOLUTIONS routine. Using current estimates of ionic concentrations, the ionic strength (U) is calculated [Eq. (A-32)]. New values of the activity coefficients G1, G2, and G3 are then calculated [Eq. (A-33)], using an intermediate variable FU. Control returns to the SOLUTIONS routine at line 470.

SINGLE VALUE RUN routine (lines 700–799)

700   Displays OUTPUT TO PRINTER Y/N prompt. Either Y or N is stored in string AN$.

710   User enters base-saturation value (BS) between 0 and 1.

720   Utilizing a temporary variable (KX), the lime potential [Eq. (A-33)] is calculated and stored as variable LP.

730   The print divider (string NP$) is displayed, followed by the log of the current values of KA and KS (i.e., the values of log K_{Al} and the ion exchange selection coefficient K_s).

750   The SOLUTIONS subroutine at line 400 is called. Using the base-saturation value (BS) entered at line 710, solutions are calculated and the results displayed.

760   If the user selected the OUTPUT TO PRINTER routine at line 700, string AN$ will contain a Y, and the program will continue to line 765. If the contents of AN$ are other than Y, the program skips to line 795.

765   The printer is turned on and set to the 80-column format.

770–775   The $\log_{10}$ of the current values of KA and KS, that is, the values of log K_{Al} and the ion exchange selection coefficient, K_s, are printed. The column headings (string N$) and the solutions for the current base-saturation value that were calculated in the SOLUTIONS subroutine and stored in string LL$ are printed. The print divider string NP$ string is then printed.

780   Printer is turned off.

795   The prompt ENTER C FOR CHECK ROUTINE

ENTER M FOR MENU

is displayed. If *M* is entered, the program returns to the menu at line 35.

CHECK routine (lines 800–900)

800–825   The total concentration (μeq/L) for both cations and anions for the most recent solutions are calculated and displayed.

830   The log K_{Al} value is calculated using the Al^{3+} and H^+ concentrations and the activity coefficients from the most recent solutions and displayed along with the input value. Here, and subsequently in this routine, the temporary variable W is used for calculation and printing.

835–840   The $\log_{10}$ values for the second and third aluminum constants [Eqs. (A-5) and (A-6), respectively] are calculated using the solution concentrations and activity coefficients as calculated by the most recent iteration. These are then displayed along with the values used internally in the program for these constants (-5.02 and -9.30, respectively).

845   Using concentrations and activities from the most recent iteration, the log of the constant for the formation of H^+ and HCO_3^- from CO_2

	and H_2O is calculated and displayed along with the internal value of -7.81 [Eq. (A-11)].
850	The left and right sides of the Gaines-Thomas aluminum–calcium relationship [Eq. (A-7)] are calculated and displayed.
900	The prompt ENTER ANY KEY FOR MENU is displayed. Program returns to menu when any key is pressed.

END (line 1000)

1000	Program ends when menu option No. 6, END, is selected.

SPECIAL VERSIONS

Listings are provided in Sect. A.4 for the portions of the program that are modified for two special versions (i.e., the SO4 SERIES version and the CO2 SERIES version). Because the changes required in the two versions are very similar, only the SO4 SERIES version will be discussed. Other than modification of the menu display to reflect the changed used of menu option No. 2, program changes are all in what was previously the SINGLE VALUE RUN routine in lines 700–799. Only those lines that have been modified or added to the original version will be discussed.

36	Menu display has been altered to reflect the changed use of option No. 2.
735	DATA statement consists of a list of the SO_4^{2-} concentrations (μeq/L) used in the series.
740	The program returns to the READ statement at line 745. An error will result when the READ statement (line 745) is encountered after all entries have been read. The ON ERR GOTO statement (line 795) used here will result in the program skipping out of the loop to the line 795 statement when this (or any other) error is encountered.
745	The next SO_4^{2-} concentration in the series is read, converted to mol/L, and stored in variable SA.
785	The program returns to line 745 to read the next DATA value.
790	This statement cancels the ON ERR GOTO command in line 740 so that any errors encountered will stop program execution and display the error message in the normal manner.

A.4 Program Listings

```
]LISTO,99

20        HOME : PRINT "BASIC EQUILIBRIUM MODEL": PRINT "SYNTHESIS
          VERSION": PRINT "MAY 1984": PRINT "J O REUSS"
21        PRINT : PRINT "PRESS ANY KEY FOR MENU": GET AN$
25        DIM A(100,12), B(20),B$(15)
30        D$ = CHR$ (4):DD$ = CHR$ (9)
35        HOME : PRINT "************* MENU *************"
36        PRINT : PRINT "1.  INPUT DATA": PRINT : PRINT "2.  SINGLE
          VALUE RUN (CHECK OPTION)": PRINT : PRINT "3.  BASE SATURATION
          SERIES": PRINT : PRINT "4.  PRINT OUTPUT": PRINT : PRINT "5.
          OUTPUT TO DISK": PRINT : PRINT "6.  END"
37        PRINT : PRINT "ENTER SELECTION": INPUT IS: ON IS GOTO
          50,700,100,40,42,1000
40        GOSUB 200
41        GOTO 35
42        GOSUB 250
45        GOTO 35
50        T1 = 10:T2 = 100:T3 = 1000:T6 = 1E6:T8 = 1E8
51        CK$ = "1"
52        HOME : PRINT
55        PRINT : PRINT : PRINT "ENTER CONCENTRATIONS IN UEQ/L": PRINT
          : PRINT : INPUT "SO4 = ";SA: PRINT : PRINT : INPUT "CL =
          ";CL:SA = SA / T6:CL = CL / T6
56        SA = SA / 2:LT = LOG (10)
60        PRINT : PRINT : PRINT "ENTER CO2 IN PERCENT": PRINT : INPUT
          "CO2 = ";CO:CO = CO / T2
65        PRINT : PRINT : INPUT "ENTER LOG AT SELECTIVITY COEFF = ";KS:
          PRINT
      : PRINT : INPUT "ENTER LOG KAL = ";KA:KS = 10 ^ KS:KA = 10 ^ KA
70        IR = 0:G1 = 0.95:G2 = 0.85:G3 = 0.75:TP = 6:PH = TP:H = 10 ^ (- PH)
          / G1
72        NP$ = "**********************************************************
          *******************"
75        NA$ = " BS    PH    H+    CA    ":NB$ + "  AL3+   AL2+   AL1+   HCO3
          ":NC$ = "  CL    SO4    ALK    CO2%"
76        N$ = NA$ + NB$ + NC$
77        A$ = "                   ':U$ = "CONCENTRATIONS ARE MICRO-EQ/L"
78        B$(1) = "BS":B$(2) = "PH":B$(3) = "H+":B$(4) = "CA":B$(5) =
          "AL3+":B$(6) = "AL2+":B$(7) = "AL1+":B$(8) = "HCO3":B$(9) =
          "CL":B$(10) = "SO4":B$(11) = "ALK":B$(12) = "CO2%"
80        DEF FN F(H) = 3 * A * H ^ 4 + 2 * AD * H ^ 3 + (1 + C) * H * H - H *
          (2 * SA + CL) - KC
85        DEF FN FP(H) = 12 * A * H ^ 3 + 6 * AD * H * H + 2 * (1 + C) * H - 2
          * SA - CL
90        GOTO 35


]LIST100,299

100   IR = 0
102   HOME : PRINT: PRINT "ENTER 1 FOR DIRECT TO DISK ": GET ID$
103   IF ID$ = CK$ THEN GOSUB 250
104   PRINT : PRINT " ENTER 1 FOR DIRECT TO PRINT ": GET IP$
105   IA = 0.9:IB = 0.18:IC = 0.03:ID = 0
110   FOR BS = IA TO IB STEP - IC
120   GOSUB 400
```

```
125  NEXT BS
130  IF ID > 0.5 THEN GOTO 140
135  IA = 0.175:IB = 0.055:IC = 0.015:ID = 1
136  GOTO 110
140  IF ID = 2 THEN GOTO 160
145  IA = 0.050:IB = 0.004:IC = 0.005:ID = 2
150  GOTO 110
160  IF ID$ = CK$ THEN GOSUB 260
165  IF IP$ = CK$ THEN GOSUB 200
170  ID$ = "":IP$ = "": GOTO 35
200  GOTO 220
201  PRINT '
220  PRINT : PRINT D$:"PR#1": PRINT
225  PRINT "LOG K SELECTIVITY = "; LOG (KS) / LOG(10);"        LOG K
     AL = "; LOG (KA)/ LOG (10): PRINT
226  PRINT U$: PRINT N$
230  FOR I = 1 to IR
235  L$ = ""
236  B$ = STR$ (A(I,1)) + A$:L$ = L$ + LEFT$ (B$,6)
237  B$ = STR$ (A(I,2)) + A$:L$ = L$ + LEFT$ (B$,6)
238  FOR J = 3 TO 11:B$ = STR$ (A(I,J)) + A$:L$ + L$ + LEFT$ (B$,7):
     NEXT J
239  B$ = STR$ (A(I,12)) + A$:L$ = L$ + LEFT$ (B$,5)
240  PRINT L$: NEXT I
245  PRINT : PRINT D$;"PR# O": PRINT : PRINT
246  RETURN
250  HOME : PRINT : PRINT "ENTER NAME OF STORAGE FILE": PRINT : INPUT F$
253  IF ID$ = CK$ THEN GOTO 295
250  PRINT D$;"MON,C,I,O"
265  PRINT D$;"OPEN ";F$: PRINT D$;:"DELETE ";F$: PRINT D$;"OPEN ";F$:
     PRINT D$;"WRITE ";F$
270  PRINT IR: PRINT 12
272  FOR J = 1 TO 12: PRINT B$(J): NEXT J
275  FOR I = 1 TO IR
280  FOR J = 1 TO 12: PRINT A(I,J): NEXT J
285  NEXT I
290  PRINT D$;"CLOSE ";F$
295  PRINT D$;"NOMON,C,I,O"
299  RETURN


]LIST300,699

400  GOTO 405
405  IR = IR + 1
410  A = (KA * G1 ^ 3) / G3:B = 10 ^ ( - 5.02) * G1 ^ 2 * KA / G2:C = 10
     ^ (- 9.30) * KA:KC = CO * 10 ^ ( - 7.81) / (G1 * G1)
415  D = (KS * BS ^ 3 * G3 * G3) / ((G2 ^ 3) * (1 - BS)Λ2):AD = A ^ (2
     / 3) * D ^ (1 / 3) + B
420  FC = FN F(H):FS = FN FP(H): IF FS < = 0 THEN H = 5 * H: GOTO 410
425  FI = FC - FS * H:FH = H
430  H = FI / ( - FS): IF H < 0 THEN H = 10 ^ ( - 7)
440  IF ABS ((FH - H) / H) < H / 100 THEN GOTO 500
450  A1 = C * H:A2 = B * H * H:A3 = A * H ^ 3
460  CA = A3 ^ (2 / 3) * D ^ (1 / 3):BC = KC / H
465  GOSUB 600
470  GOTO 410
500  PH = ( - 1) * ( LOG (H * G1) / ( LOG (10)))
510  B(1) = INT (BS * T3 + .5) / T3:B(2) = INT (PH * T3 + .5) / T3
```

```
520   B(3) = H:B(4) = 2 * CA:B(5) = 3 * A3:B(6) = 2 * A2:B(7) = A1:B(8) =
      BC:B(9) = CL:B(10) = 2 * SA:B(12) = CO * T2
525   B(11) = B(8) - B(7) - B(6) - B(5) - B(3)
530   FOR J = 3 TO 11:B(J) = INT (B(J) * T8 + .5) / T2: NEXT J
540   L$ = "":LL$ = ""
545   BS = STR$ (B(1)) + A$:L$ = L$ + LEFT$ (B$,6)
546   BS = STR$ (B(2)) + A$:L$ = L$ + LEFT$ (B$,6)
547   FOR J = 3 to 4:B$ = STR$ (B(J)) + A$:L$ = L$ + LEFT$ (B$,7): NEXT J
548   PRINT : PRINT NA$$: PRINT L$: PRINT :LL$: = LL$ + L$:L$ = ""
550   FOR J = 5 TO 8:B$ = STR$ (B(J)) + A$:L$ = L$ + LEFT$ (B$,7): NEXT J
552   PRINT NB$: PRINT L$: PRINT :LL$ + LL$ + L$:L$ = ""
555   FOR J = 9 TO 11:B$ = STR$ (B(J)) + A$:L$ = L$ + LEFT$ (B$,7): NEXT J
556   B$ = STR$ (B(12)) + A$:L$ = L$ + LEFT$ (B$,5)
560   PRINT NC$: PRINT L$: PRINT :LL$ = LL$ + L$
565   PRINT NP$
570   FOR J = 1 TO 12:A(IR,J) = B(J): NEXT J
575   RETURN
600   REM ACTIVITY COEFFICIENT SUBROUTINE
610   U = 0.5 * (H + A1 + CL + BC + 4 * (CA + A2 + SA) + 9 * A3)
615   FU = (U ^ .5 / (1 + U ^ .5) - .3 * U) * (- .509)
620   G1 = 10 ^ FU:G2 = 10 ^ (4 * FU):G3 = 10 ^ (9 * FU)
630   RETURN


]LIST700,

700   PRINT "OUTPUT TO PRINTER Y/N": GET AN$
710   PRINT : PRINT "ENTER BASE SATURATION VALUE ": PRINT "BETWEEN 0.01 A
      ND 0.90 ": INPUT BS
720   KX = ( LOG (KS * BS ^ 3 / (1 - BS) ^ 2)) / ( LOG (10) * 6):LP = KX
      + LOG (KAL)/ ( LOG (10 * 3)
730   PRINT NP$: PRINT : PRINT "KAL = "; LOG (KA) / LOG (10);"
      KS = "; LOG (KS) / LOG (10): PRINT "LIMEPOT = ";LP
750   GOSUB 400
760   IF AN$ > < "Y" THEN GOTO 795
765   PRINT DS$;"PR#1": PRINT DD$:"8ON"
770   PRINT : PRINT "KAL = "; LOG (KA) / LOG (10);"      KS = "; LOG (KS)
      / LOG (10);"        LIMEPOT = ";LP
775   PRINT : PRINT NS$: PRINT LL$: PRINT NP$
780   PRINT D$;"PR#0"
795   PRINT "ENTER C FOR CHECK ROUTINE": PRINT "ENTER M FOR MENU": GET AN
      $: IF AN$ = "M" THEN GOTO 35
800   GOTO 825
825   PRINT : PRINT "TOTAL ANIONS = TOTAL CATIONS UEQ/L":TA = (BC + CL +
      2 * SA) * T6:TC = (3 * A3 + 2 * A2 + A1 + H + 2 * CA) * T6: PRINT T
      A;" =";TC
830   W = A3 * G3 / (G1 * H) ^ 3: PRINT : PRINT "KA ENTERED = " LOG (W) /
      LT;" KA CHECK = " LOG (KA) / LT
835   W = A2 * G2 / ((G1 * H) ^ 2 * KA): PRINT : PRINT "LOG 2ND CONST
      -5.02 = "; LOG (W) / LT
840   W = A1 / (H * KA): PRINT : PRINT "LOG 3RD CONST -9.30 = " LOG (W) /
      LT
845   W = G1 ^ 2 * H * BC / CO:W = LOG (W) / LOG (10): PRINT : PRINT "CO
      2 CONSTANT -7.81 = ";W
850   PRINT : PRINT "GAINES THOMAS RIGHT SIDE = LEFT SIDE ": PRINT KS *
      (A3 * G3) ^ 2 / (VA * G2) ^ 3;" = "; (1 - BS) ^ 2 / BS ^ 3
900   PRINT "ENTER ANY KEY FOR MENU": GET AN$: GOTO 35
1000  END
```

SO4 SERIES VERSION

]LIST36

```
36    PRINT : PRINT "1.  INPUT DATA": PRINT : PRINT "2.  SULFATE SERIES
      (CHECK OPTION)": PRINT : PRINT "3.  BASE SATURATION SERIES": PRINT
      : PRINT '4.  PRINT OUTPUT": PRINT : PRINT "5.  OUTPUT TO DISK":
      PRINT : PRINT "6.  END'
```

]LIST700,799

```
700   PRINT "OUTPUT TO PRINTER Y/N": GET AN$
710   PRINT : PRINT "ENTER BASE SATURATION VALUE ": PRINT "BETWEEN 0.01
      AND 0.90 ": INPUT BS
720   KX = ( LOG (KS * BS ^ 3 / (1 - BS) ^ 2)) / ( LOG (10) * 6):LP = KX
      + LOG (KAL) / (LOG (10) * 3)
730   PRINT  NP$: PRINT : PRINT "KAL = "; LOG (KA) / LOG (10);"
      KS = '; LOG (KS) / LOG (10): PRINT "LIMEPOT = ";LP
735   DATA 0,25,50,75,100,125,150,175,200,225,250,275,300,325,350,375,
      400,425,250,475,500
740   ONERR GOTO790
745   READ SA:SA = SA / (2 * T6)
750   GOSUB 400
760   IF AN$ > < "Y" THEN GOTO 785
765   PRINT D$:"PR#1": PRINT DD$;"8ON"
770   PRINT : PRINT "KAL = "; LOG (KA) / LOG (10);"       KS = "; LOG
      (KS) / LOG (10);"        LIMEPOT = ";LP
775   PRINT : PRINT N$: PRINT LL$: PRINT NP$
780   PRINT D$;"PR#O"
785   GOTO 745
790   RESTORE : POKE 216,0
795   PRINT "ENTER C FOR CHECK ROUTINE": PRINT "ENTER M FOR MENU": GET
      AN$: IF AN$ = "M" THEN GOTO 35
```

Index